THE VIRTUES OF SUSTAINABILITY

The Virtues: Multidisciplinary Perspectives

Series Editor
Nancy E. Snow
Professor of Philosophy and Director of the Institute for the Study of Human Flourishing, University of Oklahoma

Justice
Edited by Mark LeBar

Humility
Edited by Jennifer Cole Wright

Integrity, Honesty, and Truth Seeking
Edited by Christian B. Miller and Ryan West

The Virtues of Sustainability
Edited by Jason Kawall

THE VIRTUES OF SUSTAINABILITY

Edited by Jason Kawall

OXFORD
UNIVERSITY PRESS

OXFORD
UNIVERSITY PRESS

Oxford University Press is a department of the University of Oxford. It furthers the University's objective of excellence in research, scholarship, and education by publishing worldwide. Oxford is a registered trade mark of Oxford University Press in the UK and certain other countries.

Published in the United States of America by Oxford University Press
198 Madison Avenue, New York, NY 10016, United States of America.

Library of Congress Cataloging-in-Publication Data
Names: Kawall, Jason, editor.
Title: The virtues of sustainability / edited by Jason Kawall.
Description: New York : Oxford University Press, 2021. |
Series: The virtues : multidisciplinary perspectives |
Includes bibliographical references and index.
Identifiers: LCCN 2020033688 (print) | LCCN 2020033689 (ebook) |
ISBN 9780190919818 (hb) | ISBN 9780190919825 (paperback) |
ISBN 9780190919849 (epub) | ISBN 9780190919832 | ISBN 9780190919856
Subjects: LCSH: Sustainability—Moral and ethical aspects. | Environmental ethics.
Classification: LCC GE196 .V57 2021 (print) | LCC GE196 (ebook) |
DDC 179/.1—dc23
LC record available at https://lccn.loc.gov/2020033688
LC ebook record available at https://lccn.loc.gov/2020033689

DOI: 10.1093/oso/9780190919818.001.0001

CONTENTS

SERIES EDITOR'S FOREWORD

Typically, having a virtue means being disposed to having certain kinds of perceptions, thoughts, motives, emotions, and ways one is inclined to act. The end of the twentieth and the beginning of the twenty-first centuries have seen an upsurge of interest in the topic of virtue. This is true not only in philosophy but also in a variety of other disciplines, such as theology, law, economics, psychology, and anthropology, to name a few. The study of virtue within disciplines is vitally important, yet the premise of this series is that the study of virtue in general, as well as of specific virtues, can be enhanced if scholars take into account work being done in disciplines other than their own.

Cross-disciplinary work can be challenging. Scholars trained in one field with its unique vocabulary and methods do not always move seamlessly into another discipline and often feel unqualified to undertake the task of serious cross-disciplinary engagement. The upshot can be that practitioners of disciplines can become "siloed"—trapped within their own disciplines and hesitant to engage seriously with others, even on important topics of mutual interest.

This series seeks to break the silos, with fifteen volumes on specific virtues or clusters of virtues. For each book, an introduction by the editor highlights the unity of writings by identifying common themes, threads, and ideas. In each volume, the editor seeks to include a chapter from a "wild card" discipline, a field one would not expect to see included in a collection of essays on a particular virtue. We do this both to highlight the diversity of fields in the study of specific virtues and to surprise and challenge readers to broaden their horizons in thinking about virtue.

The audience for this series is practitioners of different disciplines who seek to expand their thinking about virtue. Each volume contains chapters that are accessible and of interest to scholars from many disciplines. Though the volumes are not comprehensive overviews of the work on virtue that is occurring in any given field, they provide a useful introduction meant to pique the curiosity of readers and spur further engagement with other disciplines.

Nancy E. Snow,
Professor of Philosophy and Director of the
Institute for the Study of Human Flourishing,
University of Oklahoma

ACKNOWLEDGMENTS

I would like to express my deep gratitude to Nancy Snow, the series editor, for first suggesting the possibility of this volume many—all too many!—years ago. Nancy has provided tremendous support and thoughtful advice throughout the entire project; I greatly appreciate all of her help. Special thanks, too, to Lucy Randall at Oxford University Press for her thoughtful, informed (and informative) advice and support from the project's earliest stages. Many thanks to Hannah Doyle at OUP for her help with production, and to Leslie Anglin for copyediting. Finally, I would like to thank Heather Battaly, Babo LeMonstre, Darcia Narvaez, Thelo Rax, and the anonymous reviewers of the proposal and manuscript for their many helpful comments and suggestions.

I owe special debts of gratitude to each of the wonderful contributors to this volume. It's been a pleasure to watch these essays and the volume as whole come together, and I'm deeply appreciative of all their efforts in contributing such excellent, thoughtful pieces. I should add, too, that I greatly appreciate their patience in dealing with sometimes rather extended periods of radio silence from their editor—followed, of course, by bursts of frenzied activity! Thanks, also, for their many helpful suggestions for the collection and editing process.

I would like to thank my friends and colleagues at Colgate University—especially those in the Philosophy department and Environmental Studies program for creating such a congenial and collegial environment. Many thanks, in particular, to Lei Mara and Tracy Piatti for such excellent administrative support. And I am grateful to the Marion Hoeflich Memorial Endowment for Advancements in Philosophy for generous support for professional, expert assistance as the manuscript entered into the final stages of preparation.

Finally, for more general support, sound advice, chaos, and many helpful discussions, I am deeply grateful to Claire Sigsworth.

I dedicate this volume to my parents, Desmond and Monica, with love and gratitude.

CONTRIBUTORS

Susan Clayton is Whitmore-Williams Professor of Psychology at the College of Wooster in Ohio. She has written or edited five books, including *Conservation Psychology: Understanding and Promoting Human Care for Nature* (with Gene Myers; Wiley, 2nd edition, 2015). Her PhD, from Yale University, is in social psychology. Her research focuses on the human relationship with nature, how it is socially constructed, and how it can be utilized to promote environmental concern; she is particularly interested in applying psychological constructs of justice and identity to an environmental context. Clayton is on the editorial boards of the *Journal of Environmental Psychology* and *Social Justice Research*, and she is a fellow of the American Psychological Association, the Society for Environmental, Population, and Conservation Psychology (SEPCP), the Society for Personality and Social Psychology (SPSP), and the Society for the Psychological Study of Social Issues (SPSSI). She is a past president of SEPCP and SPSSI.

Victor Corral-Verdugo is Professor of Psychology at the University of Sonora, Mexico. He has published numerous books, papers, and chapters on environmental psychology topics, including articles in *Environment and Behavior, Environmental Management, Journal of Cross-Cultural Psychology, Journal of Environmental Psychology,* and *The Journal of Environmental Education*. His research interests include the study of sustainable behavior and its determinants, the psychological causes and consequences of climate change, and the influence of positive environments on human well-being, among others.

Christine J. Cuomo is a philosopher, feminist theorist, and feminist theorist of European descent who has worked on several collaborative projects at the intersections of environmental and social justice. She has also written and lectured widely on climate justice, ecofeminism, and other topics. From 2005 to 2015 Chris was part of a team studying landscape changes on Alaska's North Slope Borough with the help of community elders, and she is currently collaborating on research highlighting the needs and concerns of women on the front lines of the international plastic recycling industry. A professor of philosophy and women's and gender studies at the University of Georgia, her first book, *Feminism and Ecological Communities: An Ethic of Flourishing,* foregrounds the insights of ecofeminist activism, pragmatism, and multicultural radical feminism in articulating environmental values.

Matt Ferkany is Associate Professor in the Department of Philosophy and affiliated faculty of the Environmental Science and Policy Program at Michigan State University. He has also held a position in a college of education and been a fellow at the Jubilee Centre for Character and Virtue at the University of Birmingham in Birmingham, England. His teaching and scholarship focus on normative ethical problems relating to moral education, well-being and virtue, and environmental ethics and education. His work has been funded by the Spencer Foundation and published in journals such as *Pacific Philosophical Quarterly, Environmental Values, Journal of Philosophy of Education, Educational Theory,* and *Theory and Research in Education.*

Martha Frías-Armenta is Professor of Psychology and Law at The University of Sonora and a member of Mexico's National Researcher System. She has conducted research on determinants of environmental crime, family violence, alternative justice, and juvenile delinquency. Her academic production includes books, papers, and book chapters on those topics, with her articles appearing in such journals as *Environment and Behavior, Environment, Development, and Sustainability, Environmental Management, Journal of Environmental Psychology,* and *The Journal Forensic Psychiatry and Psychology.*

Cheryl Hall is an Associate Professor in the School of Interdisciplinary Global Studies at the University of South Florida. She is a political theorist with specialties in environmental political theory, feminist theory, and emotion in politics. Her research focuses on thinking through the roles that human capacities and practices might play in encouraging more just and sustainable ways of life. She is currently especially interested in questions of how to encourage political action on climate change by cultivating the emotional resources necessary to sustain cultural attention to the challenge. She is the author of *The Trouble With Passion: Political Theory Beyond the Reign of Reason* and a coeditor of *The Oxford Handbook of Environmental Political Theory.* Her articles and book chapters include "What Will It Mean to Be Green? Envisioning Positive Possibilities Without Dismissing Loss," "Framing and Nudging for a Greener Future," and "Caring to be Green: The Importance of Love for Environmental Integrity."

Laura M. Hartman is Assistant Professor of Environmental Studies at Roanoke College. Her research interests include environmental, virtue, narrative, and feminist ethics, particularly as these apply to consumption, climate engineering, and transportation. She has published articles on environmental ethics in journals such as the *Journal of Religious Ethics, Worldviews, Environmental Ethics,* and *Ethics and International Affairs.* She is the author of *The Christian Consumer: Living Faithfully in a Fragile World* (Oxford, 2011) and editor of *Flourishing: Comparative Religious Environmental Ethics* (Oxford, 2018).

Pankaj Jain is the Dean of Vijaybhoomi School of Arts and Humanities at Vijaybhoomi University. Previously, he was co-chair of the India Initiatives Group and Associate Professor in the Department of Philosophy and Religion at the University of North Texas. His recent books are *Dharma in America: A Short History of Hindu-Jain Diaspora* (Routledge, 2019) and *Dharma and Ecology of Hindu Communities: Sustenance and Sustainability* (Routledge, 2016). Currently, he is translating *Jain Darshan,* a monumental text on the philosophy of Jainism, and is working on his fourth monograph, *Makers of Modern Jainism.* He tweets at @ProfPankajJain.

Jason Kawall is Carl Benton Straub '58 Endowed Chair in Culture and the Environment, and Professor of Philosophy and Environmental Studies at Colgate University. His current research focuses on the role of knowledge (including its absence and its pursuit) in morally good and sustainable lives. More generally, his research explores virtue ethics and epistemology, with a particular emphasis on their application to environmental and sustainability issues. He has published extensively in these and related areas, with his work appearing in such journals as *American Philosophical Quarterly, Canadian Journal of Philosophy, Environmental Ethics, Ethics, Policy and Environment,* and *Philosophical Studies,* and in a number of edited volumes, including the *Oxford Handbook of Environmental Ethics* (2016) and the *Oxford Handbook of Virtue* (2018).

Anais Ortiz-Valdez is Professor of Psychology at s at the University of Sonora, Mexico. Her master's thesis addressed the relationship between human virtues and sustainable behavior, while in her doctoral dissertation she studied the moderating influence of environmental predictability/stability on the relationship between virtues and delinquent behavior. She has published papers in environmental psychology, including articles in *Environment and Behavior; Environment, Development and Sustainability,* and *Psyecology*.

Steve Vanderheiden is Professor of Political Science and Environmental Studies at the University of Colorado at Boulder. In addition to numerous articles and chapters on topics at the intersection of political theory, applied ethics, and environmental politics, he is the author of *Atmospheric Justice: A Political Theory of Climate Change* (Oxford, 2008), which won the Harold and Margaret Sprout Award for best book in international environmental politics from the International Studies Association in 2009. His current research projects focus upon the role of responsibility in response to climate change, relationships between territorial rights and carbon sinks in the context of climate mitigation efforts, and the development of democratic processes and equity norms in water governance.

Sarah Wright is Associate Professor of Philosophy at the University of Georgia. Her research focuses on the normative aspects of epistemology, particularly on the epistemic virtues as they might be developed using the Stoic moral theory as a model. She has also written on contextualism in epistemology, on social and group epistemology, and on environmental ethics. She is currently working on applications of virtue epistemology to particular real-life situations, including concerns with epistemic injustice and how the epistemic virtues can help us to avoid committing it. Her work has been published in numerous edited volumes as well as in *Episteme, Philosophical Issues, Acta Analytica, History of Philosophy Quarterly, Ethics and the Environment,* and *Metaphilosophy.*

INTRODUCTION

JASON KAWALL

As the introduction to this volume is being written, the attention of the world is being drawn to critical issues of sustainability and climate change by the teenaged Swedish activist Greta Thunberg. The school climate strikes she has helped to champion, and the multitude of related events inspired by her, are drawing millions of people together to demonstrate for urgent climate action.[1] There is a growing recognition of the potentially catastrophic impacts of human actions on future generations, and the devastating environmental harms and injustices that are already occurring; the phenomenon of global climate change, in particular, is serving to galvanize a generation of activists.[2]

1. For more on the climate strike movement, see Fridays for Future (n.d.).
2. As this final manuscript is submitted, the world is confronting the COVID-19 pandemic and its dire consequences. Still, even now, many are considering how changing norms with respect to work, travel, and other facets of our lives during the pandemic may one day help us in the pursuit of sustainability.

Jason Kawall, *Introduction* In: *The Virtues of Sustainability.* Edited by: Jason Kawall, Oxford University Press (2021). © Oxford University Press. DOI: 10.1093/oso/9780190919818.001.0001

What does the urgent call for a sustainable future mean for us—for our ways of life, and our understandings of human flourishing? What demands are placed on us as we strive to achieve such a future, and what kind of people will we need to be? What kinds of character and virtues will help us, and how can we cultivate these traits?

This volume, one of a series devoted to the interdisciplinary study of the virtues, helps to address such questions by providing a careful, wide-ranging exploration of the virtues of sustainability, and the key factors supporting their development. It brings together a diverse set of authors from a broad range of disciplines—from philosophy and political science, to religious studies and psychology. The primary focus of the volume is on the virtues and related matters of character and identity with respect to the pursuit of sustainability, rather than delving into detailed analyses or debates concerning sustainability itself, as such. That said, we begin this introduction with an overview of sustainability (and certain key issues and debates surrounding it) to provide an initial background for the reader, before turning to the virtues themselves, and introducing the chapters that follow.

SUSTAINABILITY

While the broad concept of sustainability has achieved a striking prominence in public discourse in recent years, scores of variant definitions of this core concept have been proposed by academics, governmental agencies, nongovernmental organizations (NGOs), and businesses around the world, reflecting their varying needs and purposes. Still, while many accounts have been proposed, the vast majority share a fundamental recognition that the ongoing well-being of human communities, including our social and economic systems, is intimately interwoven with the natural and ecological systems of this planet. How then, might we ensure a habitable, welcoming world for both current and future generations?[3]

Sustainability is often treated in terms of three "pillars," corresponding to environmental, social, and economic systems. The environmental dimension of sustainability concerns the maintenance or improvement of ecosystem health and services, protection of species and habitat, and a more general concern to limit deleterious effects of human activities on the rest of the natural world. In turn, the social dimension encourages us to strive for social equality, the promotion of well-functioning social structures, and the elimination of environmental, social, and other injustices, with an eye toward achieving stable, secure, and resilient communities. Finally, the economic dimension of sustainability requires economic systems that can provide adequate employment, technological innovation, and productive capacity to support and improve human well-being, as well as environmental and social sustainability, in an ongoing fashion across future generations.

Proponents of the three-pillars approach stress that we must recognize both the importance of each pillar and their interdependence. For example, there may be instances where preventing the extinction of a species is technically viable but at great financial cost, or requiring massive sacrifices to the way of life, culture, and economic well-being of an already impoverished local community—a common concern with past efforts to create pristine ecological reserves or protected areas that excluded much needed, traditional activities of local communities. As Sarah Wright notes in her chapter in this volume, proponents of the three-pillars approach would strive to avoid undermining any one of the pillars for the sake of the others, instead

3. For an excellent general introduction to sustainability, see Mulligan (2014). For a more critical discussion, drawing attention to the many different (and sometimes conflicting) accounts and definitions of sustainability, see Farley and Smith (2013).

encouraging us to find solutions that embrace the importance of all three. In the current case, proponents of the view might recommend an investment in ecotourism as a potential way forward—bringing economic benefits and stability to a local community, while encouraging the community's ongoing investment in, and concern for, the species and its habitat.[4] And where such answers are not possible, we must be keenly aware of the costs and seek balance in the future where viable.

Still, many question the prominence of the "economic" pillar in this approach, instead arguing that it should be understood as strictly ancillary, a mere means of providing resources and support for human well-being, and environmental and social sustainability. Opponents hold that the ecological and social pillars represent systems and individuals with independent, intrinsic value—values that we ought to seek to further for their own sake. The economic system lacks such inherent value. Accordingly, these opponents argue that priority should be given to the social and ecological dimensions of sustainability in decision-making, with economic concerns having only a supplemental, instrumental role in the promotion of the other two.

Similarly, while it is common in many discussions to treat the concepts of sustainability and sustainable development as essentially interchangeable, others argue that sustainable development is a deeply problematic notion. Critics worry that the concept implicitly assumes and encourages endless growth on a finite planet—a suggestion they find inherently problematic.[5] Relatedly, they worry that the

4. Of course, critiques of ecotourism *qua* green, sustainable development abound. See, for example, Duffy (2013). It is beyond the scope of this chapter to fully address these issues.

5. Proponents of sustainable development will often note in response that economic growth need not involve destructive, consumptive use of resources. For example, we can imagine an economy growing as more and more novels are written, more concerts are being performed, and people are exchanging more services.

notion of sustainable development relies on a problematic assumption that the most effective (or even the only way) to improve human well-being is through increasing consumption and growing economies. In turn, critics worry that if we do not carefully distinguish between the two concepts, we will lose sight of a genuine, desirable sustainability, and will instead be convinced to embrace a business-as-usual, flawed ideal of sustainable development.[6]

Accounts of sustainability are also commonly treated as ranging on a spectrum from weak to strong, where the strength of a given conception is a function of how readily it allows for the substitution of natural goods and resources by alternatives. For example, we might wonder whether it could be appropriately sustainable to drain a region of swamps and wetlands (which provide water filtration services, help to alleviate flooding, serve as habitat for myriad nonhuman populations, and so forth) if this allows us to expand subsidized housing developments, build new factories and hospitals, and generate additional goods and services. Is such substitution of goods sustainable?

The weakest conceptions of sustainability would hold that substitutes can be provided for any natural good, resource, or service, and that human well-being is the exclusive measure of sustainability. On such views we could, in principle, live on a planet that lacks all other life and natural systems, so long as we could generate manufactured alternatives that could equally well maintain human well-being indefinitely over generations to come.

At the other extreme, the strongest conceptions of sustainability would hold both that there is a vast range of nonsubstitutable natural goods, resources, and services, and that many of these entities have interests and a well-being of their own that must be protected and sustained.

6. For a critical overview of the three-pillars approach to sustainability and sustainable development, see Purvis et. al. (2019).

Proponents of strong sustainability often emphasize that while we can readily, fully replace much manufactured capital (we can replace a factory or a school with a more-or-less identical structure), we often have no way to truly replace various natural goods and resources—to replace a species, or to replicate the pollination work done by bees and other insects, for example. The most demanding versions of sustainability would reject almost all human-imposed changes to ecosystems and species extinctions, to the extent that we would risk losing irreplaceable goods. Such views would, of course, greatly restrict human activity. But they may also risk essentially locking in place only current (or idealized past) ecosystems and species—after all, any change leading to a new state could be construed as the loss of an irreplaceable good for which there can be no adequate substitutes.

As might be expected, most accounts lie between these extremes—recognizing that we should not be too quick to assume that we can simply find, for example, a technological alternative to the pollination work done by bees, or that nonhuman entities cannot have a morally relevant intrinsic value of their own, but also recognizing that we cannot, and should not, try to preserve all current ecosystems, species, and environments in their current state, indefinitely, and at the expense of all other goals, goods, or aspirations. These accounts encourage us to find ways forward that secure the ongoing well-being of human communities and individuals while also appropriately respecting the many values possessed by nonhuman entities and communities.[7]

A related, emerging issue concerns the extent to which humans can or ought to intervene in ecological systems for the sake of sustainability. We noted earlier that the strongest conceptions of sustainability might require us to essentially freeze the world in place, to avoid losing irreplaceable natural goods. In light of the extensive,

7. For a recent critical discussion of strong sustainability, see DesRoches (2019).

quite possibly unavoidable human impacts already occurring globally (reflected, for example, in growing suggestions that we now live in the Anthropocene age), some scholars are urging a greater acceptance of human-imposed change and intervention in ecological systems. Thus, we might wonder about the embrace of novel ecosystems—roughly, ones significantly shaped and altered by human activities and quite unlike those historically prevalent in an area, but which are now stable and self-organizing without human management. Others have wondered about the moral status of assisted migration—helping to move populations of nonhuman species to new ecosystems and habitats as their former ranges are altered or lost through climate change, habitat fragmentation, and other pressures. Proponents of such views and efforts argue that a more pragmatic orientation to environmental sustainability may be required—recognizing that novel ecosystems can still provide important ecosystem services for humans while serving as valuable habitat for nonhuman populations, and acknowledging that we may risk losing vast numbers of species if we are unwilling to intervene to save them through assisted migration or other measures (see Standish et al. [2013], and Minteer and Collins [2010] for discussion). There are thus difficult questions concerning what it is to pursue environmental sustainability in light of climate change and other anthropogenic impacts, and the extent to which we should embrace interventionist approaches to such efforts.

Finally, in considering how best to achieve and maintain sustainability, an emphasis on social and political systems would seem to be warranted, given the magnitude and scale of the challenges we face. Scholars working in the area often stress the importance of such collective and political actions, and argue that often too much emphasis is placed on small-scale, individualistic efforts that are (comparatively) ineffectual—riding bicycles to work, recycling a little more household waste, and so on. The policy initiatives, treaties, improvements to infrastructure, and other changes needed to effectively address climate change and sustainability require large-scale, governmental and institutional action.

Such institutional efforts are clearly essential, and the authors in this volume readily acknowledge their importance. But this is only half the story. Cheryl Hall and Steve Vanderheiden, in their respective chapters, capture this point well. There are limits to how much can be done through top-down, institutionally imposed changes if the general population is unwilling to abide by them. We may also wonder why political and business leaders would embrace sustainability initiatives unless they were being pressured or encouraged to act—unless there was significant public support, or they themselves were deeply invested in the importance of such efforts. In the end, we need effective, engaged leaders, and citizens who will advocate for and abide by viable efforts to seek a sustainable future.

It is worth returning to the example of Greta Thunberg. Change is often driven and inspired by individuals—supported by others, of course—but we need individuals to lead organizations, to model new ways of flourishing, and to inspire and support others. We need individuals to care about those around them, to be open-minded to new ways of life and flourishing, and to be willing to act together for the sake of a viable future. Individuals, and individual character, are essential to the pursuit of sustainability.

THE VIRTUES AND SUSTAINABILITY

With an overview of sustainability in hand, we can turn to the virtues, and their potential role in achieving and maintaining sustainability. Our discussion here will be rather more cursory, as these topics are the primary focus of the volume as a whole and will receive much attention throughout. Still, some initial groundwork is in order.

We can begin with a characterization of the virtues. Rosalind Hursthouse and Glenn Pettigrove provide an attractive account, one that would be embraced by most scholars working on the virtues and across most ethical traditions:

> A virtue is an excellent trait of character. It is a disposition, well entrenched in its possessor—something that, as we say, goes all the way down, unlike a habit such as being a tea-drinker—to notice, expect, value, feel, desire, choose, act, and react in certain characteristic ways. To possess a virtue is to be a certain sort of person with a certain complex mindset. (Hursthouse and Pettigrove, 2018, p. 4)

For example, an honest person will tend, other things being equal, to tell the truth when asked, to seek out the truth where she can, and to be careful to avoid deception, including self-deception. She will be frustrated or disappointed with dishonesty of others—and herself, should she fall short. She deeply values the truth and this is reflected in her actions, attitudes, and desires—in the ways she interprets and interacts with the world. This is an ongoing facet of who she is, her habits, and way of life. Her honesty helps to define who she is as a person.

Virtue ethics,[8] with deep roots in a wide range of philosophical and religious traditions, can be understood as being especially concerned with such questions as "How ought I to live?," "What is a good life?," and "What kind of person ought I to be?" There is thus an emphasis on examining an individual's life and character as a whole—the traits, attitudes, and habits that shape her outlook on the world, the actions she performs, and the life she leads. The virtue-ethical tradition has

8. Here taken to refer broadly to those approaches to normative ethics that emphasize the importance and centrality of the virtues and related matters of character in arriving at a robust, rich, and realistic understanding of human ethics.

re-emerged in recent decades as an attractive alternative to the deontological and consequentialist theories that have long held sway in the Western ethical tradition.

With its focus on character and the development of good habits, attitudes, and ways of life, virtue ethics is especially well-suited to addressing the moral challenges we face in the pursuit of sustainability. Much of the discourse on sustainability exhorts us to embrace ways of life (both as individuals and as communities) that will enable us to lead flourishing lives without undermining the ability of future generations to similarly flourish. And there is often an emphasis on a need to change our values, habits, and behaviors to make this possible—for example, paying attention to environmental issues that we now ignore, taking more explicit account of our impacts on future generations, becoming more politically active, consuming less and more thoughtfully, and so on. The exploration of such issues of habit, virtue, and human flourishing—and understanding the institutions and social structures that can support them—is particularly well-supported in the virtue-ethical tradition.

The virtues of sustainability can be understood, rather simply, as those virtues that will play an especially important role in allowing us to pursue and lead sustainable, flourishing lives. While most all virtues will play some role in this pursuit, it also seems clear that some virtues may take on an especially important role. For example, humility can help us to overcome hubris with respect to the human ability to control or dominate nature, or problematic anthropocentric assumptions about what has intrinsic value; it can help us to see ourselves as truly embedded in, and dependent upon, the natural systems of this planet. A virtue of simplicity would encourage reflective, thoughtful decisions with respect to consumption—what sorts of goods do we really need? Can we find better, less consumption-driven ways of life? And, of course, there will be many other plausible virtues of sustainability; in this volume we will examine several of them in depth.

THE CHAPTERS

This volume is divided into three parts. Part I, "Cultivating Virtues of Sustainability," consists of four chapters focusing on the key factors (including social and environmental structures) that support and foster the cultivation of the virtues of sustainability. Part II, "Virtues, Sustainability, and Worldviews," presents distinctive sets of virtues embraced by Jain and Iñupiat communities, with careful attention to how these virtues are manifested by the members of these various communities, and the underlying worldviews and traditions that structure and support them. Finally, in Part III, "Virtues of Sustainability," the volume turns to a set of four chapters providing in-depth discussions of particular virtues of sustainability.

Environmental psychologist Susan Clayton begins her chapter by considering whether people in fact treat sustainability and sustainable behaviors as virtuous; evidence suggests that a majority of people do, with upbringing and community consensus having significant impacts on individual beliefs. This is an important finding, as psychologists have found that those who treat sustainability or issues such as climate change as moral issues are much more likely to feel personal responsibility for helping to address them, and to perform more pro-environmental behaviors. Furthermore, differing underlying values can shape the moral attitudes of different individuals toward sustainability. Thus, those who tend toward more liberal views might be moved by appeals to justice or avoiding harm to nature. Those with more conservative tendencies could also be moved to embrace sustainability as a moral issue—but by appeal to such values as loyalty or a respect for authority.[9]

9. For more general discussion of such common differences in the moral values of liberals and conservatives, see, for example, Haidt (2012).

Clayton argues that more fundamental than virtues to the pursuit of sustainability are environmental identities. She explains that "[a] person with a strong environmental identity is described as having a perception of himself or herself as connected to or interdependent with nature. It is associated with greater concern about the environment as well as with assigning greater moral value to the environment and environmental entities." Clayton notes that identities—relatively stable senses of self that develop over time—are deeply relational, being profoundly shaped by formative experiences and our identifications with our families and communities.

Clayton then turns to the importance of family upbringing on cultivating environmental identities, noting that children are shaped by the values and attitudes of their families. But in addition, the attitudes of adults can be shaped by thinking of the future, their children, and their legacies—how they will impact and be regarded by future generations. She further argues that experiences in natural environments are important to the development of a strong environmental identity, and they may encourage more pro-social behaviors in other contexts. Clayton ends her chapter by reminding us that merely exhorting people to be virtuous may not be as effective as creating conditions that allow them to cultivate and develop strong environmental identities from which further virtues may follow.

Victor Corral-Verdugo, Martha Frías-Armenta, and Anais Ortiz-Valdez also explore the conditions that encourage the cultivation of the virtues of sustainability (and related traits) through the lens of environmental psychology. They frame their discussion by noting a puzzling but common phenomenon. While there are many character strengths associated with important virtues, and these character strengths (and virtues) are significantly associated with sustainable behavior, it is also the case that these same strengths can be used for antisocial purposes. For example, creativity can be used for malevolent purposes, and violent criminals can exhibit persistence. What then might explain the difference—why do character strengths so often contribute to sustainable behaviors, and what instead leads some individuals to apply these strengths to morally problematic or antisocial activities?

Corral-Verdugo and his coauthors note that people can be led to engage in sustainable behaviors by negative psychological factors, such as feelings of guilt or emotional discomfort when they fail to engage in such behaviors. Positive psychological factors also often play a role in producing sustainable behavior—such things as an affinity toward nature, possessing environmental skills and competencies, and feelings of satisfaction or self-efficacy can all encourage sustainable behaviors.

But Corral-Verdugo, Frías-Armenta, and Ortiz-Valdez focus on the key role of positive environments in determining how character strengths are deployed, and whether individuals are led to pro-social and pro-sustainability behaviors, and the cultivation of virtues. "In a positive environment, people see their needs satisfied, receive resources to thrive and successfully face risks and challenges; they flourish as persons and experience well-being and happiness." Under such circumstances, individuals feel compelled to protect natural and social resources; pro-social and sustainable behaviors are encouraged. Such environments provide a stability and relative calm that are enabling conditions for the cultivation of virtues. Even when adverse events occur in such environments, the community and individuals are well-placed to respond successfully. However, in the absence of such conditions—when there is uncertainty, instability, and inadequate social and environmental resources—there is greater risk that character strengths will not be manifested in pro-sustainable behaviors.

Matt Ferkany continues this exploration of the factors and conditions that facilitate the development of sustainability virtues by considering the role of schools and formal education. He focuses in particular on the potential role of the virtues in education for sustainable development (ESD), a highly prominent recent educational model promoted through UNESCO's *Decade of Education for Sustainable Development* and other influential organizations. ESD aims to influence and inform students' beliefs, values, and attitudes throughout the educational curriculum. As Ferkany notes, the goals of ESD seem to be well-captured in terms of the cultivation of sustainability virtues in students—a form of Aristotelian character education. That is, it seems that ESD is plausibly understood as attempting to cultivate certain key virtues and traits in students. Ferkany draws attention to such virtues as (1) those of cosmopolitan justice, (2) those of right attachment to people, places, and things, and (3) those of proper testimonial credulity,[10] as being especially central to the goals of ESD.

Ferkany argues that treating ESD in terms of Aristotelian virtue education will provide many benefits to ESD. For example, it will help to give shape to the curriculum; Aristotelians emphasize such factors as modeling exemplars, having opportunities to engage in and practice virtuous actions and activities, and engaging in reflection on the virtues themselves. More generally, Ferkany suggests that approaching ESD from an Aristotelian virtue education (AVE) perspective brings coherence to, and helps to elaborate the various aspects of, ESD. He argues that "specific ESD knowledge, values, and behaviors can be captured in terms of virtue. Denominating them in virtue terms should also enrich our understanding of what is educationally required to achieve them though schooling." He further argues that AVE provides ESD with a more robust framework for attitude and behavior education, holding that deep attitude change requires sustained experience of the relevant goods, and that behavior

10. Broadly, these latter are intellectual virtues that shape an individual's inquiries, enabling her to seek out and give appropriate weight to expert and other reliable testimony, while avoiding or downplaying dishonest or misleading (putative) sources of information, and so on.

change requires coupling knowledge to habits and skills that have to be acquired through routine practice in a supporting environment. Finally, Ferkany proposes that AVE would help to highlight the moral dimensions of ESD—thereby drawing attention to the relevance of the social sciences and the humanities to the pursuit of sustainability.

Still, Ferkany ends on a cautionary note, acknowledging that there are many challenges to implementing ESD, even from an AVE perspective. In particular, there are widespread misunderstandings of—and thus opposition to—character education in the United States and elsewhere. In addition, there are significant institutional, bureaucratic obstacles to implementing ESD and AVE. Moreover, we face a broader society that undercuts and undermines many of the goals of ESD. Much work will need to be done to make the approach viable and effective.

In the final chapter of Part I, Cheryl Hall notes that governmental and other institutions can nudge, impose policy, and otherwise encourage sustainable behaviors among citizens and consumers, and that these efforts are very much needed. But we also need politicians and civil servants who are invested in sustainable values, along with at least a large portion of the general population; otherwise, it is hard to envisage the impetus for effective climate action and policy. Furthermore, it is hard to impose top-down change without an open and willing population who will advocate for and embrace these values. The less people are invested in sustainability, the more regulation, policing, and further interventions will be needed. What, then, can drive such a much-needed internal motivation toward sustainability?

Hall here turns to the role of the emotions in the virtues. Virtues, according to Hall, involve a nexus of emotion, cognition, and action. And while many theorists acknowledge the role of emotions in the virtues, all too often the emotions are underplayed or largely ignored in their discussions. Hall illustrates the importance of the emotions by drawing our attention to virtues of sustainability that are characterized primarily by emotional dispositions, such as a reverence for life, or compassion for the suffering of humans and nonhumans. More generally, she argues that all virtues are at least partly constituted by emotional dispositions. For example, with respect to creativity we might find a joy or playfulness in exploring ideas, and an openness to inspiration; forgiveness will require addressing or letting go of feelings of bitterness, and so on. Hall further argues that a facility and skill in cultivating our emotions serves as a background, facilitating condition for all virtues.

Hall elaborates upon this last point by arguing that we can develop certain skills with respect to our emotions; emotions are not simply involuntary occurrences that happen to us, over which we have no control. She notes, drawing on recent empirical work, that we "encourage, discourage, channel, sublimate, project, soothe, allow, deny, reinforce, repress, intensify, label, intellectualize, and in many other ways cultivate and manage our emotions." We, of course, do not have absolute control over our emotions, but are not helpless either. We can learn to manage our emotions, developing skills to shape them at crucial moments. For example, we can train ourselves to interpret what we might construe as anxiety as anticipation or determination instead. We can learn to change our mood and emotions by shaping our environment—by playing our favorite music, or going for a walk, and so on. More generally, we can reflect on our emotions, come to understand what tends to trigger our excesses, and identify and question the beliefs that lie behind them; such emotional work is a critical aspect of virtue cultivation.

In Part II of the volume, we turn to two studies of sustainability virtues as embodied in particular communities and traditions. Pankaj Jain focuses on virtues of sustainability emerging from the Jain tradition—a tradition where, he argues, philosophy, religion, and environmentalism are interwoven and ultimately inseparable. Jain introduces broad features of Jain cosmology and illuminates how moral guidance arises within this broader metaphysical context. He focuses in particular on three important Jain virtues: *ahimsā* (nonviolence), *aparigraha* (nonattachment or nonacquisitiveness), and *anekāntavāda* (pluralism). He illustrates how these virtues are manifested in both the lives of ordinary Jain community members, as well as those of renowned leaders, monks, and other exemplars.

Jain notes that while India faces severe sustainability challenges, it is also a nation whose citizens have (1) small ecological footprints and (2) many moral resources to ground a concern for nature—and that these important advantages can often be traced to the influence of the Jain and Buddhist traditions. In particular, the low-impact, vegetarian diet of most Jains (and other Indians through their influence) can be seen as especially impactful at the level of individual households. He notes that even as such dietary practices are just now gaining widespread popularity across the West, the Jain community has embraced them for centuries as part of a more holistic, embodied tradition and way of life.

Finally, throughout his discussion, Jain emphasizes the crucial role of exemplary individuals in Jainism. From the revered early Jain teacher and reformer Mahavira, to Mohandas Gandhi, Jain laypeople often strive to model themselves and their behaviors upon the actions and writings of the great exemplars and leaders of their tradition, as well as those who have been deeply influenced by it. Jain demonstrates that there is much to be learned from the Jain community for those seeking models and exemplars of sustainable ways of life, while carefully addressing the objections of those more skeptical of this view.

Christine J. Cuomo focuses on the values and virtues of northern, indigenous Iñupiat communities, with a particular emphasis on their core virtue of respect for nature. Cuomo notes that, as with many indigenous communities and cultures, the Iñupiat embrace a rich, biocentric worldview that profoundly shapes their values and ways of being in the world. For the Iñupiat there is a sense of relationship with the land, animals, plants, and other entities that informs their respect for nature. This latter virtue is manifested in such activities as learning from and communicating with nature, and in respectful harvesting (refraining from greed and waste). Moreover, this respect shapes and converges with other important values and virtues of Iñupiat communities.

Cuomo draws attention to the tensions and challenges as these communities work to articulate and abide by their long-standing values in the face of vast oil and gas extraction developments. As Cuomo notes, the challenges imposed by such developments have in fact often led to these communities reconnecting with and revitalizing their long-standing values. She writes of Alaska's North Slope that it is "the ancestral home of an indigenous community that has maintained and actualized commitments to show respect for nature through subsistence hunting, harvesting, traveling, camping, and practicing spiritual culture in a place where oil and gas leases are active, seismic activity and toxic dumping are commonplace, and the land is changing, eroding, and suffering due to climate change."

Recent work in environmental and sustainability ethics in the Western tradition often suggests that we need whole new sets of virtues and radical revisions of ethics as a whole to cope with unprecedented challenges of climate change and sustainability. Indeed, some academic Western philosophers have pointed to a respect for nature as a virtue in recent decades. But as Cuomo notes, such discussions tend to treat respect for nature as a radical, possibly controversial, and unprecedented virtue. There is little or no attention paid to other traditions—particularly indigenous traditions—where a respect for nature has been lived and deeply embedded within communities for generations as a fundamental virtue. And all too often, what attention has been paid to such communities is abrupt and dismissive.

Cuomo, drawing on work by Thom White Wolf Fassett, argues that environmental ethics, rather than requiring radically new values and virtues, should instead be understood in terms of rediscovery, and of recognizing more basic values and virtues from a variety of cultures and traditions. But this must be an active engagement and rediscovery, involving critical reflection, creativity, and ongoing discourse, rather than a mere attempted return to a mythologized or romanticized past.

Cuomo's chapter, with its extended exploration of respect for nature, serves as an effective transition to Part III of the volume. Here we turn to in-depth discussions of particular virtues of sustainability, beginning with Laura M. Hartman's careful articulation and exploration of cooperativeness. Hartman draws on a rich body of scholarship in the Christian ethical tradition to argue that cooperativeness will be essential to the pursuit and maintenance of sustainability; from individuals to NGOs to national governments, we must find ways to work together, to overcome past differences, and to pursue shared goals.

Hartman begins by drawing attention to a key distinction between cooperation—a matter of individual actions, which can be for good or for evil, and cooperativeness—a virtue that manifests itself in cooperation for the good, with the good. She illuminates the distinction (in part) by discussing the difficulties that many in wealthy nations face in avoiding cooperation or complicity in evil, drawing on work by the theologian Julie Hanlon Rubio. For example, buying clothing made in sweatshops, driving personal vehicles, and many other common activities could plausibly be understood as being complicit in, or cooperating with, evil. But Hartman also notes the significance of seeking opportunities to cooperate with the good—perhaps in buying used clothing from a charity shop or organization.

Hartman examines cooperativeness's close connections to such virtues as patience and forgiveness, and she illustrates a broad range of circumstances in which it will play a key role in achieving and maintaining sustainability, particularly in its social and environmental dimensions. Hartman explores ecological restoration and cooperatives as particularly compelling examples of cooperativeness in the context of sustainability. She closes by noting the challenges of achieving and manifesting a genuinely virtuous cooperativeness, drawing attention to cases where well-intended, often admirable individuals and communities have been led astray into deeply flawed forms of cooperativeness, or (in some cases) even cooperating with evil.

Jason Kawall turns to patience as a potential virtue of sustainability, noting that two very different attitudes toward this possibility seem likely. On one hand, it may seem that patience will be invaluable to our pursuit of sustainability, given an emphasis on the gradual achievement of large-scale, long-term change, where impacts may take decades to become manifest. On the other hand, it may instead seem that we cannot afford patience, given the need for drastic, immediate action by individuals, governments, and industry. Kawall suggests that with a plausible understanding of patience as a virtue, the first view—that it is a key virtue of sustainability—is correct.

To defend this claim, Kawall presents an understanding of the virtue of patience that combines aspects of two recent, influential accounts by Matthew Pianalto and Nic Bommarito. He then draws on this combined account to illustrate that patience will be a crucial virtue in pursuing and maintaining sustainability across a variety of contexts—from engaging in ongoing political and social action (and avoiding activist burnout), to responding appropriately to setbacks and mistakes, to developing new habits and ways of life.

Kawall argues that a focus on patience in contexts of sustainability can help us to recognize often-overlooked aspects of the virtue more generally. For example, in addressing such issues as climate change, we require patience (1) in pursuing goals where we ourselves will never see the full impacts and outcomes of our actions (which may only be felt decades or hundreds of years later), and (2) in continuing the pursuit of our long-term goals and actions despite a lack of clear, short-term or immediate feedback. Kawall argues that this latter concern, in particular, will be one of the greatest challenges we will face, and it will arise in a variety of contexts. Patience and perspective will be essential.

Steve Vanderheiden's starting point lies in the recognition that humanity as a whole, and particularly those leading wealthy, consumerist lifestyles, are having unsustainable impacts upon the natural world. This fact can be captured through a variety of measures, from ecological footprints, to estimates of how many Earths it would take to sustain a given lifestyle (for a given global population). He argues that knowledge of such facts should help to ground a virtue of conscientiousness, where such a virtue would entail (1) an awareness of our impacts, our embeddedness in (and fundamental dependence upon) natural systems, and (2) corresponding normatively laden attitudes and actions shaped by this awareness.

Vanderheiden compares conscientiousness to various civic virtues—all can be understood as virtues directed toward sustaining the well-being of a larger community. He notes that in the case of conscientiousness, this means a concern both to protect environmental goods and values but also to engage positively in relationship with the broader ecological community. Conscientiousness can foster both a global concern for a cosmopolitan, global community, but through its emphasis on our embeddedness, also a localized concern for particular communities and places.

Vanderheiden argues that a virtue-ethical approach of this kind can better capture a rich, plausible ethics for sustainability, including climate ethics, than other prominent, rival ethical approaches. For example, while a consequentialist ethical outlook might suggest that we can lead highly environmentally demanding, consumption-laden lifestyles so long as we purchase a corresponding set of ecological (e.g., carbon) offsets, a virtue-ethical approach would have us look more broadly at questions of our wastefulness, the requirements of a good life, and potential injustices in allowing the wealthy to simply continue to pollute as before. More generally, Vanderheiden argues that even while broader, systematic changes in policy and technology will be needed to achieve sustainability, we must not ignore the crucial role of individual character. Unless individuals embrace various virtues of sustainability, politicians, CEOs, and other powerful individuals will have little reason to embrace and act upon sustainability concerns.

Sarah Wright, in the concluding chapter of the volume, draws attention to the importance of epistemic environmental virtues—virtues that drive and enable our search for truth and understanding. She focuses in particular on the virtues of creativity and open-mindedness as being of particular importance to the pursuit of sustainability.

Wright argues that we will need creativity to develop new technologies, policies, and ways of life in our pursuit of sustainability. Such creativity can be exhibited in small day-to-day actions and localized inventions or policies, but also in projects at far larger scales; the overall, cumulative of effect of such wide-ranging creative projects will be essential to a global, enduring sustainability. Wright addresses common worries concerning creativity as a virtue—in particular, that it is not a character trait that can be cultivated and developed (as we would expect with genuine character virtues), but is instead simply a gift that a few lucky individuals happen to be born with. She also addresses the concern that creativity is not an epistemic virtue insofar as highly creative ideas may in fact drift very far from reality—to wild hypotheses and overly imaginative, impractical proposals. In both cases, Wright draws on work in empirical psychology and the broad virtue-ethical tradition to undermine these concerns.

Wright treats open-mindedness as something of a complementary virtue to creativity. It is all well and good to develop highly creative and effective solutions to problems, but this will be of little use if those who would need to implement such solutions cannot or will not accept them. Ideally, of course, we will generate creative proposals (for technologies, policies, and the like) that are sensitive to community needs, traditions, and ways of life. But individuals and communities must also, in turn, demonstrate an open-mindedness and willingness to learn and adapt. We need to be open to hearing new ideas, to learning valuable truths, and to embracing appropriate new technologies, policies, and ways of life. It is hard to imagine us being successful in any of these ventures without creativity and open-mindedness.

CONCLUSION

There is still much to be considered as we strive toward sustainability. As noted earlier, significant work remains to be done even in clarifying the goals and nature of sustainability. But on any plausible such account we will face significant challenges, and our character—be we politicians, CEOs, or just plain citizens—will be put to the test.

It is our hope that this volume will serve to inspire and provide a foundation for further work at the intersection of character, virtue, and sustainability. Lessons concerning the place of familial and social environments, educational systems, and the importance of the emotions and virtuous role models can be applied, developed, and further explored. The current volume may also provide additional impetus to rediscover and rethink familiar and neglected virtues from across a range of traditions and sources. Finally, and at the same time, we hope that it may encourage further consideration of whether there may be genuinely new potential virtues, habits, and ways of life to be discovered that could aid us as we pursue an enduring, global sustainability.[11]

REFERENCES

DesRoches, C.T. 2019. "Some truths don't matter: The case of strong sustainability." *Ethics, Policy & Environment*, 22(2), 1–13.

Duffy, R. 2013. *A Trip Too Far: Ecotourism, Politics and Exploitation*. New York: Routledge.

Farley, H.M., and Z.A. Smith. 2013. *Sustainability: If It's Everything, Is It Nothing?* New York: Routledge.

Fridays for Future. n.d. "Fridays for future." Accessed April 20, 2020. https://www.fridaysforfuture.org/.

Haidt, J. 2012. *The Righteous Mind: Why Good People Are Divided by Politics and Religion*. New York: Pantheon Books.

Hursthouse, R., and G. Pettigrove. 2018. "Virtue ethics." In *The Stanford Encyclopedia of Philosophy*, edited by E.N. Zalta. https://plato.stanford.edu/archives/win2018/entries/ethics-virtue/.

Minteer, B.A., and J.P. Collins. 2010. "Move it or lose it? The ecological ethics of relocating species under climate change." *Ecological Applications*, 20(7), 1801–1804.

Mulligan, M. 2014. *An Introduction to Sustainability: Environmental, Social and Personal Perspectives*. New York: Routledge.

Purvis, B., Y. Mao, and D. Robinson. 2019. "Three pillars of sustainability: In search of conceptual origins." *Sustainability Science*, 14(3), 681–695.

Standish, R.J., A. Thompson, E.S. Higgs, and S.D. Murphy. 2013. "Concerns about novel ecosystems." In *Novel Ecosystems: Intervening in the New Ecological World Order* (1st ed.), edited by R.J. Hobbs, E.S. Higgs, and C.M. Hall, 296–309. Hoboken, NJ: Wiley-Blackwell.

11. Many thanks to an anonymous reviewer for very helpful comments on this introduction.

THE VIRTUES OF SUSTAINABILITY

PART I

CULTIVATING VIRTUES OF SUSTAINABILITY

Chapter 1

Virtue and Sustainable Behavior

A Psychological Perspective

SUSAN CLAYTON

Is sustainability virtuous? This chapter utilizes a psychological perspective to examine the links between virtue and sustainability. I start by exploring the concepts of sustainability and virtue, and how they incorporate psychological aspects. I then discuss the consequences of virtue and why it might matter if sustainability is perceived to be virtuous. Expanding on the concept of virtue, I incorporate questions of justice and examine virtues that might be particularly connected to sustainability and then describe the kind of person who might develop a virtuous approach to sustainability. Finally, I consider the possibility of a virtuous cycle, in which a healthy natural environment could make us virtuous. I close with some comments about the relevance of virtue to human behavior.

Susan Clayton, *Virtue and Sustainable Behavior* In: *The Virtues of Sustainability*. Edited by: Jason Kawall, Oxford University Press (2021). © Oxford University Press. DOI: 10.1093/oso/9780190919818.003.0001

SUSTAINABILITY, VALUES, AND VIRTUE

In its most general sense, *sustainability* refers to a system that can persist indefinitely, meeting the needs of the present while maintaining the ability to meet the needs of the future (e.g., World Commission on Economic Development, 1987). This begs the question of what is to be sustained. Sustainability is commonly defined as having three prongs: ecology, economy, and society. That is, it incorporates ecological sustainability (conserving species and ecosystems); economic sustainability (maintaining a functioning economic system in which people can make a living); and social sustainability (protecting social relations by insuring a reasonable level of equity). A sustainability perspective recognizes that resources are limited and, importantly, that resource use occurs within a system in which one person's use of resources diminishes the amount of resources available to others. Thus the term *sustainability* clearly connects environmental and social concerns, acknowledging the interdependence among ecosystems, social systems, and individual well-being.

This immediately raises another question, particularly among anyone who is working to promote the health of ecosystems, social systems, or individuals: how to make the tradeoffs that may be required when trying to foster more than one type of well-being. The need to make choices highlights the relevance of virtue. Virtue references individual character, reflecting a person who makes morally good decisions; a person who makes good decisions about trading off environmental and social well-being must have certain attributes. Practical knowledge about the impacts and unintended consequences of particular policies is one such attribute, but good decisions also require attention to values, a point that has been acknowledged by the Intergovernmental Panel on Climate Change (cf. Gardiner, 2006). A virtuous person, then, is one who makes decisions that reflect the correct values.

Values are preferences for end states or ways of being, and although most people will agree on a positive assessment of most values, such as honesty, luxury, or security, there are stable individual differences in the ways in which they are prioritized, as well as more culturally agreed-upon hierarchies of values. See, for example, Cuomo's discussion of Inupiaq values in Chapter 6 of this volume. One common distinction is made between self-enhancing values, which focus on personal interests such as power and hedonism, and transcendent values, which focus on the well-being of others beyond the self, including other people, the community, or the environment (see Corner, Markowitz, and Pidgeon, 2014). Prioritizing transcendent values over more self-interested values will presumably be considered more virtuous, perhaps in part because it reflects the strength of will to override self-centered impulses. However, choosing between altruistic (focused on other people) and biocentric (focused on the natural world) values may require more difficult decisions.

Nature is sometimes considered to have sacred value (Fiske and Tetlock, 1997), a perception that may lead people to prioritize the protection of nature and reject tradeoffs with other values, such as economic benefit. This can lead to some difficult conundrums. A deontological perspective, which emphasizes moral rules, prohibits the kind of rational comparisons between environmental and human costs that allocate value to both, and which would be advocated under a more consequentialist perspective that emphasizes the utilitarian outcome of decisions (Sacchi, Riva, Brambilla, and Grasso, 2014). In fact, many who value nature would nevertheless give higher priority to social and economic well-being. For example, a small qualitative study of environmentally engaged people found that social justice values were described as more important motivators than biospheric ones (Howell, 2013).

LINKING SUSTAINABILITY TO VIRTUE

Conceptual reasons for linking sustainability to virtue are based on the recognition of what sustainability may involve. Sustainability raises important issues about the distribution of scarce resources; about rights, entitlements, and responsibility; about inclusion and citizenship; and about the relationship between individual behavior and collective well-being. Philosophers specialize in conceptual analysis and in explicating definitions of fairness, and several philosophers have explored, for example, the ethics of climate change (Gardiner, 2006; Jamieson, 2010; see Grasso and Markowitz, 2015). Sustainable behavior may be described as virtuous because it avoids taking more than one's share of environmental resources; because it protects the rights of others to a minimum threshold of environmental benefits; because it recognizes the responsibility of individuals to promote collective well-being, perhaps including the well-being of the natural world; and because it acknowledges the existence of others in a collective community and the moral obligation to do no harm to others.

A psychological perspective, however, is more concerned with perceptions than with conceptual arguments: Do people perceive that virtue is relevant to sustainability? The empirical evidence suggests that most people do consider sustainability to have a moral component. Kempton, Boster, and Hartley (1996) identified a strong moralistic component to environmental values in the United States. In a series of studies examining people's values for the natural world, Stephen Kellert (e.g., 1996) found that the second most common set of values (as expressed in interviews and surveys) were *moralistic*: based on assigning ethical standing to nature and natural entities. More recently, Markowitz (2012) surveyed a (nonrepresentative) sample of over 900 college students and found that 40%–50% said that climate change was "an ethical or moral issue," while only 23%–29% said that it was not. Notably, Wang (2017) found that the perception that climate change was a moral issue was predicted in part by perceived social norms, showing that people look to others for help in deciding what is a moral issue. Both childhood upbringing and the consensus of a relevant community have a large impact on individual beliefs about what is virtuous.

MORAL INTUITIONS

Psychologists are also interested in the consequences of these perceptions. What difference does it make that people link sustainability to virtue? Such a linkage is important in part because of the relevance of moral intuitions. Researchers (e.g., Haidt, 2001) have shown that we have intuitive emotional responses to issues that have ethical implications. These emotional responses are rapid and likely to affect behavior unless we consciously engage in more deliberate, rational cogitation. Thus the link between sustainability and virtue has implications for environmental concern and behavior. In Markowitz's (2012) study, those who said that climate change was an ethical issue also considered it to be more serious, and they were significantly more likely to perceive personal responsibility for doing something about it and to report intentions to engage in more sustainable behavior. Other researchers have also reported that the tendency to perceive a moral aspect to environmental issues is associated with more pro-environmental attitudes (Feinberg and Willer, 2013; Wang, 2017), and that perceived moral obligations predict pro-environmental intentions and behavior (Barbarossa and De Pelsmacker, 2016). In particular, people who feel moral obligations toward a larger range of natural entities, either dispositionally or because they are induced to do so, are more likely to show pro-environmental behavior (Bratanova, Loughnan, and Gatersleben, 2012). Those who perceive climate change as a moral issue make more climate-friendly food choices (Vainio & Mäkiniemi, 2016).

Moral foundations theory (e.g., Graham et al., 2011) identifies several core moral concepts, which have clear connection to the idea of virtues. The original five concepts identified in moral foundations theory include care (vs. harm), purity, fairness, loyalty, and respect for authority, the first three of which seem clearly relevant to sustainable behavior. Moral concerns for the environment can relate to issues of care in that unsustainable behavior can impinge on the welfare of others, including nonhuman others. A content analysis of videos and newspaper op-eds (Feinberg and Willer, 2013) found that persuasive environmental messages were more likely to emphasize harm vs. care than the other four moral concepts; aversions to harm can be linked to concern for current or future generations, nonhuman animals, and anthropomorphized aspects of the environment. Research by Wang (2017) found that Americans' perception that climate change was a moral issue was predicted by perceived risks to others, but not risks to self.

Purity is also relevant to environmental messaging, in that pollution is seen as degrading the environment. Concerns for purity are linked to viewing the environment as imbued with sacred value and therefore worthy of being protected at all costs. Considerations of fairness arise when some people or groups use more than their "fair share" of resources, or when they contaminate the environment of others. Discussions of environmental justice, or of intergenerational justice, typically raise this issue, identifying as problematic situations where some groups (e.g., ethnic minorities or future generations) are left to adapt to the consequences of unsustainable behavior by others.

Several researchers have explored the impact of framing environmental concerns on the basis of alternative moral concepts. Feinberg and Willer (2013) explored the idea that environmental attitudes are based in part on a moral component. Using a diverse US sample, they found that their respondents evaluated a person who was described as behaving unsustainably (not recycling) as less moral than someone whose behavior was more sustainable, or a control condition whose sustainability was not described. However, this was only true among politically liberal respondents. Further research showed that liberals were more likely than conservatives to use moral terms to describe their environmental attitudes. The researchers believed this was due to the fact that the typical emphasis on care versus harm tends to be more stressed by liberals; conservatives, meanwhile, tend to respond more strongly to themes of purity, loyalty, and respect for authority. In an experimental manipulation, the researchers stressed either purity or harm in an environmental message. In the harm condition, and in a control condition, liberals scored higher on an environmental attitude measure than conservatives. In the purity condition, though, liberals and conservatives were not significantly different. Similarly, Wolsko (2017) found that conservation attitudes and intentions were higher among liberals than conservatives following a message that stressed harm or justice (or in a control condition), but that there was no difference associated with political orientation following a message stressing loyalty and respect for authority. This set of studies showed the relevance of moral principles to environmental attitudes and also highlighted the existence of individual differences in response to different moral issues.

WHEN MORALITY AND SUSTAINABILITY ARE NOT LINKED

Although some people recognize the moral or ethical component of environmental issues, many clearly do not, or do not act upon the recognition. Gardiner (2006) has suggested that climate change presents a "perfect moral storm": that is, its characteristic features make it difficult for people to accurately perceive the moral implications. These characteristic features include the spatial and temporal distance between cause and effect; the fact that responsibility for climate change is diffuse, not traceable to a single or small number of actors or actions; and the lack of a global system that can coordinate an effective response—all of which prohibit a clear assignment of responsibility for responding to climate change. In addition, scientific uncertainty (e.g., about the precise timing and effects of climate change as well as of potential responses) further complicates the issue. These characteristics do not make it impossible to recognize the moral relevance of sustainable behavior. However, according to Gardiner, they create a context in which moral corruption can allow us to avoid such recognition. In other words, when virtuous behavior is clear, it will be acknowledged, but when the virtuous behavior is slightly less clear, self-interest can promote alternative perspectives.

A study by Bastian, Zhang, and Moffat (2015) provided an experimental demonstration of the role of self-interest. They studied the role of moral convictions about mining in predicting acceptance of mining in three countries where it is a significant industry: Australia, Chile, and China. Although moral convictions against mining were related to reduced acceptance of mining in all three countries, the effect of moral conviction on attitudes was weaker when there were perceived economic rewards from mining. In general, inequity in vulnerability to environmental changes means that self-interest will have different implications for different actors (individuals, nations, and other interest groups) motivating different ways of thinking about the situation and different perspectives on appropriate actions. Selective attention, in particular, can be deployed in order to avoid recognizing the morally virtuous course of action.

JUSTICE AND SUSTAINABILITY

Determinations of right and wrong behavior are relevant because resource limits require decisions to be made. If there were enough environmental resources for everyone to have all they needed, there would be no need to look for principles to guide the use and allocation of those resources. The recognition that environmental resources are becoming scarce, because the supply is either limited (e.g., fossil fuels), becoming degraded by pollution (e.g., water), or shrinking due to a changing climate (arable land), means that decision rules must be developed to trade off competing claims. Do people who have a history with a place deserve greater access to its resources, or should resources be distributed equally? Do those who contributed most to an environmental problem, such as climate change, have an obligation to compensate those who are bearing the costs? Even when people have accepted the focus on an interdependent community that is inherent in the idea of sustainability, the requirements for inclusion in this community remain crucially undetermined. There will be disagreement over whether other species, including plants or insects, have moral standing (e.g., Opotow, 1993), or whether future generations have the same entitlement to consideration as those currently alive (Syme, Kals, Nancarrow, and Montada, 2000). Sustainability mandates a consideration of the future, but it does not necessarily assign it the same value as the present.

The perceptions of what is fair affect people's willingness to engage in sustainable behavior. Sweetman and Whitmarsh (2016) found that perceived fairness was a significant predictor of willingness to undertake both personal and political actions on behalf of the environment, and a more powerful predictor than perceived efficacy or personal identity. Interestingly, seeing fellow group members engage in such behaviors could influence judgments of perceived fairness. Sweetman and Whitmarsh argue that high-status group members have the ability to shape moral norms that will in turn influence the rest of the group. The implications for sustainability could be depressing, given the lack of sustainable behavior that characterizes most high-status members of Western societies. However, Schuldt and colleagues (Schuldt, Pearson, Romero-Canyas, and Larson-Konar, 2017) took advantage of Pope Francis's message about climate change in order to test experimentally the impact of a moral role model on attitudes toward climate change among Americans. In a nationally representative sample, they found that reminding respondents of Pope Francis and his views about climate change increased agreement that climate change was a moral or ethical issue.

Perceived justice has also been shown to be an important predictor of the acceptability of environmental policies (Clayton, 2018). Dreyer and Walker (2013) examined Australian citizens' acceptance of an actual, not hypothetical, carbon policy and found that perceived fairness was an important predictor, more so than perceived effectiveness or ideology. A report on UK attitudes toward energy systems change found that respondents made reference to a number of values, including respect for nature, fairness, and social justice (Parkhill, Demski, Butler, Spence, and Pigeon, 2013). The researchers conclude that, although policy costs are important, "meaningful public acceptability [of policies] may only be achieved if it is rooted, in a significant way, in the described value system" (p. 38). A perception that environmental policies are consistent with values, including values for justice, may be important in determining policy acceptance.

Beliefs about justice can vary; for example, Sacchi et al. (2014) found that people who had a deontological orientation toward the natural environment, emphasizing that decisions about the environment should be based on moral principles, were less positive about a cap-and-trade policy (a system in which an upper limit is set on the carbon emissions of any one business or organization, but they may "trade" with other organizations that are not using their full allowance) than those who had a less deontological perspective. Significantly, perception that the cap-and-trade policy would be effective increased support for the policy, but only among those whose deontological orientation was low; when deontological orientation was high, perceptions of effectiveness did not affect attitudes toward the policy.

Opinions about environmental policy are not always consistent with rational assessment: acceptance of environmental policies is related to a perception that they are fair but also to personality variables. In research examining support for a range of environmental policies, Clayton (2018) found that ratings of the fairness of specific policies were a stronger determinant of acceptability than perceived effectiveness of the policy in general. Although liberals and environmentalists were, as expected, more supportive of environmental policies in general, endorsement of environmental justice principles outweighed the impact of these individual differences for the most part. However, political orientation and environmentalism each helped to predict attitudes toward some of the policies, even after taking endorsement of environmental justice into account. Overall, the policies that were considered most acceptable were the ones that were least restrictive on individuals, such as providing them with information about their energy use or encouraging them to reduce their use, although these weren't considered as effective as, for example, taxing their energy use or prohibiting inefficient devices. Because people are reluctant to accept constraints on their behavior, this may be demonstrating the role of self-interest in affecting judgment that Gardiner (2006) described as "moral corruption."

ENVIRONMENTAL VIRTUES

Focusing more specifically on a virtue ethics approach to sustainability implies that sustainable behavior can reflect on the moral character of the person behaving. Such an effect can be seen in a series of studies showing that people's sense of shame predicted their pro-environmental intentions (Amatulli, De Angelis, Peluso, Soscia, and Guido, 2019). Interestingly, guilt did not serve to predict intentions in the research by Amatulli et al., confirming that the behavior should be seen as reflecting on personal identity in order to affect intentions. Guilt is taken to imply a negative evaluation of one's behavior, whereas shame is considered to imply a negative evaluation of one's self. Other research, however, has found guilt to be related to the intention to take action to address climate change (Wang, 2017). It is worth noting that not all research clearly acknowledges a distinction between shame and guilt. Rees, Klug, and Bamberg (2015) found that shame and guilt in response to environmental degradation clustered together in a "guilty conscience" reaction, which in turn predicted both behavioral intention and actual behavior in support of environmental protection.

A number of virtues can be identified that have environmental relevance: for example, simplicity, moderation, and frugality. Other chapters in this volume emphasize patience (Kawall), cooperativeness (Hartman), or conscientiousness (Vanderheiden). Humility is often considered to be particularly important, because recognizing one's own limitations and the fact that humans are not the center of the universe can, in turn, encourage restraint of one's own behavior in deference to the needs of others. Significantly, humility is often cited as an aspect of significant or transcendent experiences in the natural world (Williams and Harvey, 2001).

Jordan and Kristjansson (2017) describe a number of environmental virtues, including compassion, benevolence, and temperance. They go on to highlight a new virtue, *harmony with nature*, as key to a virtue ethics approach to sustainability. Harmony with nature is defined as a perception that nature permeates all aspects of our lives, and a recognition of the interdependent relationship between humans and nature that incorporates awe, wonder, and fundamentally a sense of connection. So although it is value-based, it is also related to a cognitive construction of the world that links humans to the natural world in an egalitarian relationship, rather than situating people above or apart from the rest of nature. Jordan and Kristjansson argue that the sustainability problem is caused by a tendency to perceive the relationship between humans and nature in a way that ignores the interconnections between them. This point has a solid grounding in environmental ethics. The foundation of ethics, according to Aldo Leopold (1949), is that the individual is a member of an interdependent community, and the land ethic extends the boundaries of the community to include nature.

What kind of person is likely to be characterized by the virtue of harmony with nature? Jordan and Kristjansson (2017) describe such a person as possessing

> a way of thinking and of seeing the world that recognises the complex network of relationships that surround us, that recognises that we are part of a larger ecological system and that feels wonder and fascination towards that natural world; that acknowledges the interrelatedness of the individual, society, and environment, and that is aware of the ever-changing nature of life and therefore understands the need for continual change in themselves. (p. 1220)

Notably, this definition not only incorporates a way of understanding the world and an emotional response to that world; it also, by referencing the "need for continual change," has implications for personal development.

ENVIRONMENTAL IDENTITY

Putting this description in the context of personality, Clayton (2003, 2012) has developed the concept of *environmental identity*. A person with a strong environmental identity is described as having a perception of himself or herself as connected to or interdependent with nature. It is associated with greater concern about the environment as well as with assigning greater moral value to the environment and environmental entities. Describing the predictors of sustainability in terms of an identity rather than a virtue highlights the relational aspect, so that one's orientation toward nature fundamentally implicates one's self-concept. An identity is a relatively stable sense of self that develops over time as a result of experiences and of one's interpretation of those experiences. This leads to a consideration of the type of experiences that might develop an environmental identity.

As is true for most formative experiences, early experiences tend to have more impact than later ones. In particular, identity can be conceptualized within a family setting. Identities are fundamentally relational, describing who we are like, who we are connected to, and who we are dissimilar from (Boddy et al., 2016; Clayton, 2003). In a self-reinforcing process, experiences within family groups both capitalize on existing identities (as a member of the family) and develop the meaning of that identity. Family norms have both a descriptive and a prescriptive aspect, as children are socialized into a variety of specific behavioral practices. The relevance of family interactions in promoting sustainability has not received much attention to date, but it is significant because of what we know about the role of family in contributing to environmental identity as well as moral development.

In a qualitative approach to this topic, Boddy and colleagues (2016) interviewed (relatively well-off) families in both India and the United Kingdom to examine the moral narratives within which sustainability practices were couched. They specifically note the performative aspect of such practices, which implicitly or explicitly reflected on the moral character of those engaging in the environmentally relevant practices. Some of these families described themselves as skeptical about environmentalism, but nevertheless made a point of describing the ways in which their practices were aimed at inculcating moral values, such as awareness of the value of things and avoiding a "disposable way of life" (p. 362). Others were more explicit about trying to create an awareness of relative privilege and the responsibilities that go with it, including a responsibility to protect environmental resources through sustainable behavior. Significantly, these practices reflected on family identity by making explicit comparison with others who were not so fortunate but who were also not as informed about environmental issues, or perhaps were immoderate in their attention to them—showing an absence of the prudence that is required by a virtue ethics approach. By describing the role of the family in the development of a moral identity as well as in the creation of norms about sustainable behavior, this study illustrates the potential connection between moral identity and sustainability.

It is interesting to note the relevance of parent–child relations in activating concerns about the morality of sustainability. Anecdotally, it often seems to be the case that people become sensitized to the importance of sustainable behavior when they think about the future for their children or their grandchildren. Zaval, Markowitz, and Weber (2015) empirically demonstrated the impact of thinking about the future; their research found both correlational and experimental evidence that legacy motives were associated with pro-environmental behavior. Legacy motives were measured by items, including "I care what future generations think of me"; in a separate study, they were manipulated by having people think about the positive impact they would like to have on future generations. People who scored higher in legacy motivation, or who were experimentally prompted to consider their legacy, showed greater pro-environmental behavior intentions, and were more likely to donate money to an environmental organization. Apparently thinking about one's connection to the future makes salient the desire to present a self whose values are connected to their behavior.

Of course, families are not the only sources of moral education for children. Educational settings are also influential (see Ferkany, Chapter 3, this volume) and may be able to encourage the development of environmental virtues by providing experiences that are relevant to sustainability and presenting role models who can demonstrate virtuous engagement with the issue.

Although Clayton (2003) described a standard scale of written questions to measure environmental identity, other authors have described different ways of measuring the perceived sense of connection to nature. An important early measure, which correlates strongly with Clayton's measure, was the Inclusion of Nature in the Self scale (INS). The INS was adapted by Schultz (2001) from an earlier measure that assessed a perceived bond between people by asking people to identify which of a set of circle diagrams best described the relationship. The two circles, representing the individual and nature, ranged from complete overlap to complete lack of overlap, using a spatial metaphor for the degree of relational closeness. However, a newer measure (Martin and Czellar, 2016) was constructed to assess not only the amount of overlap but also whether a person defined his or her relationship to nature as more hierarchical or more egalitarian. Through exploratory research, they found that people's drawings of their relationship with nature varied meaningfully in the distance, size, and centrality of nature, as well as in its overlap with self. They also found that their Extended INS measure was significantly related not only to environmental behavior but also to environmental values.

This provides further evidence that conceptions of the relationship between self and nature are important for understanding the way values might play a role in guiding sustainable behavior. Some writers have suggested that an appropriate conception of oneself as small, rather than large; at the same level, rather than above; and connected, rather than isolated, is important for promoting altruistic or virtuous behavior—arguing, for example, that "action within collectives . . . requires diminished attention to the self" (Piff, Dietze, Feinberg, Stancato, and Keltner, 2015, p. 883).

THE INFLUENCE OF NATURAL ENVIRONMENTS

Experiences with natural environments are important in the development of a strong environmental identity. Research suggests that those who grow up in rural settings are more likely to have a strong environmental identity as an adult, perhaps due to social interactions that highlight the value and emotional significance of nature, as well as memories that are more strongly associated with the natural world (Prévot, Clayton, and Mathevet, 2018). We need to consider not only the potential for people to demonstrate virtue through the way in which they impact their environments by acting sustainably or unsustainably but also the potential for environments to inculcate virtue. Whereas an environmental identity develops over time, a growing body of research suggests that even short-term experience in natural environments may in fact promote virtuous behavior. Research has shown, for example, that people are more generous in rooms containing plants than in those without plants (Weinstein, Przybylski, and Ryan, 2009) and are more helpful after being exposed to more beautiful rather than less beautiful nature (Guéguen and Stefan, 2016; Zhang, Piff, Iyer, Koleva, and Keltner, 2014). Corral-Verdugo et al. (Chapter 2, this volume) rightly highlight the ways in which positive environments can encourage both human well-being and sustainable behavior.

Piff et al. (2015) suggest that certain experiences can encourage a reconceptualization of the self in a way that diminishes self-interest, what Zhang et al. (2014) refer to as "unselfing." Piff et al. argue in particular for the impact of awe, which they describe as driving a feeling that one's own interests are less significant in the face of something "vaster" than oneself. Awe is associated with a perception that one is part of a collective. Research by Piff et al. found not only that the experience of awe was associated with more ethical, prosocial, and generous behavior, but specifically that natural imagery was able to create this experience, and that the impact of awe on behavior seemed to be mediated by a perception of oneself as "small," for example, "I feel small or insignificant." In one study, they asked their participants to gaze up at a grove of towering trees (or, in a control condition, a tall building). Gazing at the trees for one minute increased ratings of awe, which in turn increased helpfulness and ethical decision making, and reduced perceived entitlement. This is consistent with other research suggesting that nature is capable of inducing "transcendent" experiences and feelings of humility (Williams and Harvey, 2001).

It may not be an overstatement to say that experiences of nature may be able to increase virtuous behavior, apparently by shifting attention away from the self. It would be only fitting if this virtuous behavior included behaving more sustainably, taking care of the nature that was responsible for the feeling of awe. Merrick (in Vining and Merrick, 2012) reports that "environmental epiphanies," or experiences in nature that shift a perception of the relationship between nature and self, were associated with long-term changes in environmentally responsible behavior.

CONCLUSION

Many influences on sustainable behavior lie outside of the individual. It is important to acknowledge the extent to which individual impact on ecosystems, for example through carbon emissions or use of environmental resources, is a function of structural factors such as economics and physical infrastructure, or cultural norms. For example, some of the most significant individual behaviors are use of private automobiles, meat eating, and childbearing: reliance on cars is profoundly affected by the structure of work and the availability of public transportation; eating meat is strongly affected by culture as well as financial resources; and the choice of whether and how many children to have is linked to financial and economic development. Because virtue reflects individual character as demonstrated through behavioral choices, it cannot be demonstrated in areas over which people have no control. Nevertheless, there are ways in which individuals have a great deal of autonomy and flexibility in determining their own impact on the environment. This gives them the opportunity to be virtuous in the way in which they make those decisions.

Research suggests that many people do consider sustainable behavior to be virtuous, though they do not necessarily act on a basis of that belief. The likelihood of sustainable behavior may be enhanced if people are reminded of its moral relevance, for example by highlighting moral values in a persuasive communication—though only if the values are personally important (Feinberg and Willer, 2013; Wolsko, 2017)—or by exposure to a moral role model (Schuldt et al., 2017; Sweetman and Whitmarsh, 2016). A heavy-handed moral frame, however, can also induce a defensive reaction (Tauber, van Zomeren, and Kutlaca, 2015).

Research also indicates that the way in which people think about their own relationship to the natural world can profoundly influence their tendency to act in ways that protect the environment. A more profound sense of connection to the natural world encourages environmental concern and pro-environmental behavior. Although this effect might be due to a greater personal significance for nature, so that it is more self-relevant and more highly valued, some research suggests that it may be affected by the sense of humility that comes from putting oneself in the context of something larger (Piff et al., 2015).

It would be comforting to conclude that perceptions of interdependence, and a self-concept that recognizes one's own limits and belonging to something greater, may be facilitated by experiences in nature and may in turn promote virtuous behavior of the kind associated with sustainability. But we need to bear in mind the fact that experiences in nature are decreasing. People are spending less time in natural settings and are even exposed to nature less often through books and film (Prévot-Julliard, Julliard, and Clayton, 2015); with increasing urbanization and a growing population, this is inevitable. Will people be less virtuous in the future? Can we find other experiences, perhaps virtual encounters with nature, that will evoke the same sense of humility? Can experiences of nature be transformed in ways that will have the same, or a similar, impact on the sense of self (Clayton et al., 2017) by encouraging people to see themselves as part of something larger?

Finally, history has shown that virtue alone is not a particularly strong motivator of human behavior. Subjectivity in definitions as well as the existence of other motives creates a context in which virtue is more aspirational than definitional. Aspirations are not irrelevant; the perception that sustainability is virtuous sets in place a prescriptive social norm that people will try to live up to when it is highlighted. However, trying to convince people to be sustainable by telling them it is virtuous may not be as effective as trying to create the conditions that lead them to see their relationship with the world in a particular way, in particular by developing an environmental identity that leads them to want to live in harmony with nature. Such an orientation may be increasingly at odds with individualistic social mores, but it may be necessary if we are to have a sustainable society.

REFERENCES

Amatulli, C., M. De Angelis, A. Peluso, I. Soscia, and G. Guido. 2019. "The effect of negative message framing on green consumption: An investigation of the role of shame." *Journal of Business Ethics, 157*, 1111–1132.

Barbarossa, C., and P. De Pelsmacker. 2016. "Positive and negative antecedents of purchasing eco-friendly products: A comparison between green and non-green consumers." *Journal of Business Ethics, 134*(2), 229–247.

Bastian, B., A. Zhang, and K. Moffat. 2015. "The interaction of economic rewards and moral convictions in predicting attitudes toward resource use." *PloS One, 10*(8), e0134863.

Boddy, J., A. Phoenix, C. Walker, U. Vennam, H. Austerberry, and M. Latha. 2016. "Telling 'moral tales'? Family narratives of responsible privilege and environmental concern in India and the UK." *Families, Relationships and Societies*, 5(3), 357–374.

Bratanova, B., S. Loughnan, and B. Gatersleben. 2012. "The moral circle as a common motivational cause of cross-situational pro-environmentalism." *European Journal of Social Psychology, 42*(5), 539–545.

Clayton, S. 2003. "Environmental identity: A conceptual and an operational definition." In *Identity and the Natural Environment*, edited by S. Clayton and S. Opotow, 45–65. Cambridge, MA: MIT Press.

Clayton, S. 2012. "Environment and identity." In *Oxford Handbook of Environmental and Conservation Psychology*, edited by S. Clayton, 164–180. New York: Oxford University Press.

Clayton, S. 2018. "The role of perceived justice, political ideology, and individual or collective framing in support for environmental policies." *Social Justice Research, 31*, 219–237.

Clayton, S., A. Colléony, P. Conversy, E. Maclouf, L. Martin, A.C. Torres, M.X. Truong, and A.C. Prévot. 2017. "Transformation of experience: Toward a new relationship with nature." *Conservation Letters, 10*(5), 645–651.

Corner, A., E. Markowitz, and N. Pidgeon. 2014. "Public engagement with climate change: The role of human values." *Wiley Interdisciplinary Reviews: Climate Change, 5*(3), 411–422.

Dreyer, S., and I. Walker. 2013. "Acceptance and support of the Australian carbon policy." *Social Justice Research, 26*(3), 343–362.

Feinberg, M., and R. Willer. 2013. "The moral roots of environmental attitudes." *Psychological Science, 24*(1), 56–62.

Fiske, A., and P. Tetlock. 1997. "Taboo trade-offs: Reactions to transactions that transgress the spheres of justice." *Political Psychology, 18*(2), 255–297.

Gardiner, S. 2006. "A perfect moral storm: Climate change, intergenerational ethics and the problem of moral corruption." *Environmental Values, 15*(3), 397–413.

Graham, J., B. Nosek, J. Haidt, R. Iyer, S. Koleva, and P. Ditto. 2011. "Mapping the moral domain." *Journal of Personality and Social Psychology, 101*(2), 366–385.

Grasso, M., and E. Markowitz. 2015. "The moral complexity of climate change and the need for a multidisciplinary perspective on climate ethics." *Climatic Change, 130*(3), 327–334.

Guéguen, N., and J. Stefan. 2016. "'Green altruism': Short immersion in natural green environments and helping behavior." *Environment and Behavior, 48*(2), 324–342.

Haidt, J. 2001. "The emotional dog and its rational tail: A social intuitionist approach to moral judgment." *Psychological Review, 108*(4), 814–834.

Howell, R.A. 2013. "It's not (just) 'the environment, stupid!' Values, motivations, and routes to engagement of people adopting lower-carbon lifestyles." *Global Environmental Change, 23*(1), 281–290.

Jamieson, D. 2010. "Climate change, responsibility, and justice." *Science and Engineering Ethics, 16*(3), 431–445.

Jordan, K., and K. Kristjánsson. 2017. "Sustainability, virtue ethics, and the virtue of harmony with nature." *Environmental Education Research, 23*(9), 1205–1229.

Kellert, S. 1996. *The Value of Life: Biological Diversity and Human Society*. Washington, DC: Island Press.

Kempton, W., J. Boster, and J. Hartley. 1996. *Environmental Values in American Culture*. Cambridge, MA: The MIT Press.

Leopold, A. 1949. *A Sand County Almanac*. New York: Oxford University Press.

Markowitz, E. 2012. "Is climate change an ethical issue? Examining young adults' beliefs about climate and morality." *Climatic Change, 114*(3), 479–495.

Martin, C., and S. Czellar. 2016. "The extended inclusion of nature in self scale." *Journal of Environmental Psychology, 47*, 181–194.

Opotow, S. 1993. "Animals and the scope of justice." *Journal of Social Issues, 49*(1), 71–85.

Parkhill, K., C. Demski, C. Butler, A. Spence, and N. Pidgeon. 2013. *Transforming the UK Energy System: Public Values, Attitudes and Acceptability—Synthesis Report*. London: UKERC.

Piff, P., P. Dietze, M. Feinberg, D.l Stancato, and D. Keltner. 2015. "Awe, the small self, and prosocial behavior." *Journal of Personality and Social Psychology, 108*(6), 883–899.

Prévot, A.C., S. Clayton, and R. Mathevet. 2018. "The relationship of childhood upbringing and university degree program to environmental identity: Experience in nature matters." *Environmental Education Research, 24*(2), 263–279.

Prévot-Julliard, A.C., R. Julliard, and S. Clayton. 2015. "Historical evidence for nature disconnection in a 70-Year time series of Disney animated films." *Public Understanding of Science, 24*(6), 672–680.

Rees, J., S. Klug, and S. Bamberg. 2015. "Guilty conscience: Motivating pro-environmental behavior by inducing negative moral emotions." *Climatic Change, 130*(3), 439–452.

Sacchi, S., P. Riva, M. Brambilla, and M. Grasso. 2014. "Moral reasoning and climate change mitigation: The deontological reaction toward the market-based approach." *Journal of Environmental Psychology, 38*, 252–261.

Schuldt, J., A. Pearson, R. Romero-Canyas, and D. Larson-Konar. 2017. "Brief exposure to Pope Francis heightens moral beliefs about climate change." *Climatic Change, 141*(2), 167–177.

Schultz, P. 2001. "The structure of environmental concern: Concern for self, other people, and the biosphere." *Journal of Environmental Psychology, 21*, 327–339.

Sweetman, J., and L. Whitmarsh. 2016. "Climate justice: High-status ingroup social models increase pro-environmental action through making actions seem more moral." *Topics in Cognitive Science, 8*(1), 196–221.

Syme, G., E. Kals, B. Nancarrow, and L. Montada. 2000. "Ecological risks and community perceptions of fairness and justice: A cross-cultural model." *Risk Analysis, 20*(6), 905–916.

Täuber, S., M. van Zomeren, and M. Kutlaca. 2015. "Should the moral core of climate issues be emphasized or downplayed in public discourse? Three ways to successfully manage the double-edged sword of moral communication." *Climatic Change, 130*(3), 453–464.

Vainio, A., and J.P. Mäkiniemi. 2016. "How are moral foundations associated with climate-friendly consumption?" *Journal of Agricultural and Environmental Ethics, 29*(2), 265–283.

Vining, J., and M. Merrick. 2012. "Environmental epiphanies: Theoretical foundations and practical applications." In *Oxford Handbook of Environmental and Conservation Psychology*, edited by S. Clayton, 485–508. New York: Oxford University Press.

Wang, X. 2017. "Risk perceptions, moral attitudes, and anticipated guilt in US consumers' climate change behavioral intentions." *Journal of Risk Research, 20*(12), 1554–1567.

Weinstein, N., A. Przybylski, and R. Ryan. 2009. "Can nature make us more caring? Effects of immersion in nature on intrinsic aspirations and generosity." *Personality and Social Psychology Bulletin*, 35(10), 1315–1329.

Williams, K., and D. Harvey. 2001. "Transcendent experience in forest environments." *Journal of Environmental Psychology, 21*(3), 249–260.

Wolsko, C. 2017. "Expanding the range of environmental values: Political orientation, moral foundations, and the common ingroup." *Journal of Environmental Psychology, 51*, 284–294.

World Commission on Environment and Development. 1987. *Our Common Future.* New York: Oxford University Press.

Zaval, L., E. Markowitz, and E. Weber. 2015. "How will I be remembered? Conserving the environment for the sake of one's legacy." *Psychological Science*, 26(2), 231–236.

Zhang, J.W., P. Piff, R. Iyer, S. Koleva, and D. Keltner. 2014. "An occasion for unselfing: Beautiful nature leads to prosociality." *Journal of Environmental Psychology*, 37, 61–72.

Chapter 2

The Environmental Psychology of Pro-Sustainability Virtues

VICTOR CORRAL-VERDUGO, MARTHA FRÍAS-ARMENTA, AND ANAIS ORTIZ-VALDEZ

INTRODUCTION

Sustainability is an approach to policy and action that intends to cope effectively with the serious environmental problems faced by humankind. These problems do not only affect the integrity of the natural environment; they also affect personal and social well-being. Such situations exhibit the intricate interdependency between humans and the physical environment. They also reveal that social problems cannot be solved without solving environmental problems—and vice versa (Vucetich and Nelson, 2010). As such, *sustainable behaviors* (SBs; i.e., actions intended at, simultaneously, meeting human needs and conserving the environment) should necessarily aim to protect both the social and the natural milieu (Tapia, Corral-Verdugo, Fraijo, and Durón, 2013).

Unfortunately, the current dominant approach to sustainable development (sometimes called "the reformist approach"; see

Victor Corral-Verdugo, Martha Frías-Armenta, and Anais Ortiz-Valdez, *The Environmental Psychology of Pro-Sustainability Virtues* In: *The Virtues of Sustainability*. Edited by: Jason Kawall, Oxford University Press (2021). © Oxford University Press. DOI: 10.1093/oso/9780190919818.003.0002

Clifton, 2012) fails to acknowledge the interconnectedness of social and environmental problems. This implies that a shift in the dominant approach is required, to promote "a change in how humans relate to each other and to nature" (Jordan and Kristjánsson, 2015, p. 1205). Virtue ethics may be an important component of this change since SBs are, presumably, of a virtuous nature (Corral-Verdugo, Tapia, and Ortiz, 2015; Hilbig, Zettler, Moshagen, and Heydasch, 2013). Virtues are seen as dispositions that are "partly constitutive of being a good person" (Sumner, 1998, p. 18), and doing right is a characteristic of pro–environmentally and pro–socially oriented individuals. Moreover, the need to extend moral consideration to nonhuman elements of nature is increasingly mentioned in academic forums, thus taking into account the individual's moral obligation to protect the nonhuman environment (Norton and Thompson, 2014). For these reasons, the study of the virtues of sustainability is an important research topic.

The field of normative ethics contains three dominant theoretical and (supposedly) competing domains addressing matters of right action: virtue ethics, consequentialism, and deontology. *Virtue ethics* focuses on virtues or character strengths, while *deontology* emphasizes duties or rules, and *consequentialism* (including utilitarianism) stresses the consequences of actions (Kawall, 2009). In the case of SB, a utilitarian will focus on the positive consequences (environmental quality, well-being) of acting pro-environmentally and pro-socially. In turn, a deontologist will emphasize acting in accordance with a moral rule ("protecting others," "preserving the environment"), and a virtue ethicist will point out the fact that being pro-sustainable is benevolent or fair (Hursthouse and Pettigrove, 2016). These three aspects have been directly or indirectly addressed in research concerning the determinants of SB. For instance, it has been demonstrated that pro-environmental and pro-social behaviors promote environmental quality (Jenks and

Jones, 2010) but also impact on well-being of society or people generally, as utilitarianism predicts. People who engage in SBs are very often motivated to do so because of their moral impulses or sense of responsibility (Rezvani, Jansson, and Bengtsson, 2017), as deontology anticipates. Fair, altruistic, and benevolent individuals tend to protect the natural environment and constantly engage in pro-social behaviors (Corral-Verdugo et al., 2015; Sevillano, Aragonés, and Schultz, 2010); moreover, their actions result in positive feelings of satisfaction, self-efficacy, psychological well-being, and happiness in the agent herself (Brown and Kasser, 2005; Corral-Verdugo, González, Rascón, and Corral-Frías, 2016), as virtue ethicists foresee. This latter aspect (the virtue ethics of sustainability) is of enormous importance in research concerning the determinants of SB.

Although the relationship between virtues and SB seems logical, virtuous traits sometimes lead to unexpected outcomes. Character strengths that are expected to be conducive to pro-environmental and pro-social actions sometimes are instead used *to harm society and the environment* (Cropley, Kaufman, and Cropley, 2008; Silke, 2004; Trinca, Ferrari, and Lees, 2008). Therefore, the issue of the virtuosity of SB is more complex than it appears. The question thus becomes, what makes people translate their virtuosity into SBs? Environmental psychology considers ways of responding to such inquiry. This field of psychology studies the reciprocal influences of environmental factors and human behavior, so that the resulting findings may be used to improve both environmental quality and human well-being (Stokols and Altman, 1987). Therefore, the environmental psychology of pro-sustainability virtues should indicate what environment–behavior interactions result in pro-sustainability virtues. In addition, it should determine how the environment influences the virtues–SB relationship so that virtues can be used for pro-social and pro-ecological purposes.

DETERMINANTS OF SUSTAINABLE BEHAVIOR

Understanding people's motivation to engage in SB helps to solve environmental problems, which explains why a growing number of social and behavioral scientists are committed to study the determinants of SB (Clayton and Saunders, 2012; Stern, 2000). One avenue of study is the investigation of *negative* psychological factors that drive pro-sustainability behavior. For instance, it is known that people practice SB to avoid fear, guilt, shame, feelings of hypocrisy, emotional discomfort, and sacrifice, produced by failing to engage in pro-environmental actions (Dickerson, Thibodeau, Aronson, and Miller, 1992; Kaiser, Schultz, Berenguer, Corral-Verdugo, and Tankha, 2008; Malott, 2010). Although realistic, this approach is incomplete, since it misses the powerful influence that *positive* psychological factors have on pro-sustainable actions. These factors include positive intrinsic motivation (satisfaction, self-efficacy, autonomy) and happiness that result from engaging in pro-environmental and pro-social behaviors (Brown and Kasser, 2005; Corral-Verdugo et al., 2016). Other positive variables involve human capacities (pro-environmental skills and competency), positive emotions (affinity toward nature), and positive personality traits (openness to experience, kindness, future orientation, character strengths, etc.) that drive the practice of SBs (Hirsch, 2010; Hsu, 2004; Kals, Schumacher, and Montada, 1999; Milfont, Wilson, and Diniz, 2012). An important aspect to consider in this approach is the influence of *positive environmental factors* that facilitate engaging in prosocial and pro-environmental activities (Baum and Gross, 2017). By considering a framework that addresses favorable situational factors and the benefits—not only the sacrifices—obtained from engaging in pro-sustainable practices, environmental psychology in accordance with the tenets of positive psychology establishes the principle that acting rightfully provides reward (which also coincides with the virtue ethical approach to normative ethics). In addition, by investigating contextual clues to pro-sustainability, a more coherent and inclusive explanation of the drivers of SB can be obtained. As Seligman and Csikszentmihalyi (2000) summarize:

> The social and behavioral sciences can play an enormously important role. They can articulate a vision of the good life that is empirically sound while being understandable and attractive. They can show what actions lead to wellbeing, to positive individuals, and to thriving communities. (p. 5)

We would amend this sentence to include *positive contexts* with actions that result in well-being, positive individuals, and thriving communities. The consideration of positive contexts represents a more inclusive approach based on the synergistic effort of environmental psychology, positive psychology, and normative ethics.

VIRTUES AND CHARACTER STRENGTHS

Morality represents more than fulfilling duties and producing the positive consequences associated with the practice of SB. Developing character strengths and acting virtuously may also enable sustainability. Virtues and character strengths are positive facets of human psychology that have been associated with SB. Virtues, it is important to note, also promote positive outcomes for oneself and one's surroundings (Han, 2015). Jordan and Kristjánsson (2015) propose that virtue ethics may help to transform the dominant approach to human–nature relations currently based on a social domination approach (humans above nature, humans above other humans). Virtue ethics would not only redress the human–nature relationship and facilitate a pro-ecological perspective; it also would promote a more holistic and integrative view of sustainability. It would encourage humanity to search for ways to live and flourish in a more sustainable world.

Sustainable activities can be categorized as virtuous since virtues are conducive to personal growth, positive coexistence with other people, happiness, and environmental adaptation (Peterson and Seligman, 2004). Moreover, virtues oppose consumerism and inequity, two major causes of unsustainability (Corral-Verdugo et al., 2015); thus, SBs are supported by character strengths (Corral-Verdugo, 2012).

Virtues are "manifestations of excellence or character strength that make it possible for individuals to flourish as human beings" (Fowers, 2005, p. 4). Particular character strengths are part of virtues, which means that strengths are under virtues within a hierarchical organization. Peterson and Seligman (2004) offer a scheme of virtues and character strengths. This scheme has served as the foundation for empirical studies of the relationship between virtues and SB. They classify six virtues and 24 character strengths, as the following list shows:

Virtues	Character Strengths
1. *Wisdom*	Creativity, curiosity, open-mindedness, love for learning, and perspective
2. *Courage*	Bravery, persistency, honesty, and zest
3. *Humanity*	Love, kindness, and social intelligence
4. *Justice*	Teamwork, equity, and leadership
5. *Temperance*	Forgiveness, humility, prudence, and self-regulation
6. *Transcendence*	Appreciation of beauty and excellence, gratitude, hope, humor, and spirituality

THE MULTIPLE FACETS OF SUSTAINABLE BEHAVIOR

Sustainability encompasses economic, cultural, productive, political, social, and behavioral dimensions, and strategies aimed at maintaining a balance between the satisfaction of human needs and the conservation of the environment (Gouveia, 2002). The behavioral dimension of sustainability addresses the role psychological factors (human cognitions, emotions, capacities, and actions) play in the induction of a pro-sustainable way of life.

The main subject of the psychology of sustainability is SB, the set of actions aimed at protecting oneself, others, and the environment (Corral-Verdugo, Frías, and García, 2010; Pato and Corral-Verdugo, 2013). SB is based on an individual's activities that sustain her or him while, at the same time, allowing the sustenance of other human beings and the conservation of the natural milieu. Therefore, SB includes self-protective, pro-social, and pro-environmental behaviors.

Self-protective behaviors are the basis of SB. An individual who is not able to protect her personal (psychological, physical) integrity would hardly be sustainable (i.e., not able to sustain herself). Self-protective behaviors include activities that promote physical and psychological health, such as eating healthy food, engaging in physical exercise, avoiding risky and addictive behaviors, and practicing actions that result in personal flourishing and sometimes in the promotion of spirituality (Bermúdez, 1999; Mahoney et al., 2005). Most self-protective behaviors have a positive impact on the social and natural environment; for instance, healthy eating local food reduces carbon emissions and decreases resource consumption. Protecting self-integrity permits individuals to pay attention to the needs of others, while achieving personal growth and flourishing lead to environmentally protective behaviors (Corral-Verdugo, Caso, and Frías, 2017), and so on.

Since sustainability includes a social component, pro-sustainable individuals also practice pro-social actions aimed at benefiting, protecting, and empowering others, especially those not genetically related to them. These actions include altruistic and equitable behaviors. Altruistic behaviors imply activities directed at selflessly providing benefits to people in need (Van Lange, 2000). Financially supporting charities; giving money, food, or clothing to the homeless; visiting sick people at hospitals; volunteering with community activities; and donating blood are examples of altruistic behaviors. In turn, equitable behaviors help empower others by promoting a fair distribution of resources and avoiding discrimination and/or a biased treatment to people, based on personal, demographic, or biological characteristics (Corral-Verdugo, García, Castro, Viramontes, and Limones, 2010). Providing equal educational opportunities for girls and boys, treating subordinates in a fair manner, and respecting social diversity are instances of equitable behavior. Equitable actions also guarantee a balance between human well-being and ecosystem integrity, facilitating access to resources for people and the preservation of the natural environment (Tapia et al., 2013).

Additional characteristics of pro-sustainability individuals include pro-ecological and frugal actions. Pro-ecological behaviors constitute actions that protect natural resources (Kaiser, Schultz, Berenguer, Corral, and Tankha, 2008; Tapia et al., 2013); they manifest as recycling, using non-fossil sources of energy, pro-environmental lobbying, and ecosystems preservation, among many others. In turn, frugal behaviors ameliorate the impact of human activities on the integrity of natural resources by decreasing unnecessary consumption of products and living lightly (Iwata, 2002). Reuse, avoiding food waste, choosing to use a bicycle or walk instead of driving a vehicle, and moderate consumption of resources are examples of frugal behavior.

Do the virtues encourage or entail SB? Or at minimum, are SBs of a virtuous nature? We will try to address these questions in the next section.

SUSTAINABLE BEHAVIOR AND VIRTUES

Evidence exists that character strengths and virtues are significantly associated with SB. *Creativity* is a component of pro-environmental competency (Fraijo, Corral-Verdugo, Tapia, and González, 2010). *Open-mindedness* helps people move toward pro-sustainable lifestyles (Lourdel, Martin, and Bérerd, 2006; Wright, Chapter 10, this volume). Individuals who persuade others to engage in pro-environmental behaviors (Corral-Verdugo, Hess, Hernández, and Suárez, 2002) exhibit *perspective*. Moreover, some authors suggest that participation in environmental protection programs stimulates student *zest* (Ballantyne, Fien, and Packer, 2001). *Love, kindness,* and *social intelligence*—character strengths of the humanity virtue—have been associated with pro-environmental and pro-social actions (Fraijo et al., 2010; Van Lange, 2000). Strengths associated with the virtue of justice are essential for SB: *teamwork* is important in environmental conservation (Hargreaves, 2010), *equity* is implicit in SB (Corral-Verdugo et al., 2010), and *leadership* has the power to drive environmental activism (Werder, 2006).

Prudence is linked to frugality, one indicator of SB (Corral-Verdugo, 2012), and *self-regulation* associates with the practice of pro-ecological actions (Pelletier, 2002). *Gratitude* toward life and nature is a determinant of environmental conservation (Corral-Verdugo, 2012), while *spirituality* promotes both pro-ecological and altruistic behaviors (Vaillant, 2008). This review implies that each virtue is implicated in the manifestation of SB, which suggests that these virtues are likely components of SB.

To test this concept, Corral-Verdugo et al. (2015) conducted a study that investigated the whole set of character strengths included in the VIA Inventory of Strengths (Peterson and Seligman, 2004) and a measure of SB encompassing pro-ecological, frugal, altruistic, and equitable actions self-reported by participants. They confirmed the research hypothesis, finding significant associations between the six virtues—including the 24 character strengths—and the four indicators of SB. In terms of specific character strengths, appreciation of beauty, creativity, perspective, and self-regulation were more strongly associated with the whole range of SB. Equitable behaviors and pro-ecological actions demonstrated associations with character strengths that were stronger than those produced by altruistic and frugal behaviors. However, the fact that all the virtues correlated significantly with the four kinds of SB led the authors to conclude that their study "could potentially provide an opportunity to expand the existing research on moral concepts in environmental psychology into an even broader scope of moral virtuousness without an extensive discussion of the conceptual discrimination (between specific virtues)" (Corral-Verdugo et al., 2015, p. 894).

Why are virtues and SBs so closely related? Today, sustainability may help to define what it means to be "good." "Being sustainable" is now seen as an important part of "being good" (Thompson, 2007), and most people aspire to be good because, by doing so, they may achieve well-being, obtain rewards, and gain social recognition. It is true that the meaning of goodness has varied significantly over time because each epoch in human history is characterized by different sets of virtues that provided guidance as to what it meant to be good. If our contemporary definition of *sustainability* delineates, at least in part, what it means to be good, we need to specify the ethical and behavioral schema that leads to sustainability.

Vucetich and Nelson (2010) argue that such a sustainability schema involves understanding how to balance three primary virtues of sustainability: concern for human needs, ecosystem health, and social justice. In addition, according to these authors, we need to establish a distinction between "vulgar" and "virtuous" sustainability, since sustainability "could mean anything from exploit as much as desired without infringing on future ability to exploit as much as desired to exploit as little as necessary to maintain a meaningful life" (Vucetich and Nelson, 2010, p. 543). Nonetheless, there is no empirical proof that "vulgar" (exploit as much as desired *without compromising the future*) and "virtuous" (exploit as little as possible) pro-sustainable practices are significantly different from each other. Nor is there evidence of the existence of the proposed three primary virtues of sustainability or their influence on SB. This lack of evidence presents an opportunity to further investigate the association between virtues and SB.

Some scholars establish additional distinctions among virtues that could explain what specific kind of virtues/character strengths lead to SB. Baehr (2011), for instance, examines the relationship between intellectual and moral virtues, assuming they are two different types of virtues. According to the author, an intellectual virtue is a trait that depends on its being oriented towards epistemic goods—"the wanting of which can be explained in terms of our natural curiosity" (Grimm, 2009, p. 245). Conversely, a moral virtue depends on being oriented toward the well-being of others (a facet of pro-sustainability orientation). It may be that moral virtues are more conducive to sustainability than intellectual virtues. Yet, although Baehr (2011) identifies a distinction between both types of virtues, he assumes that an overlap between the two classes is likely, a situation that complicates the issue of what types of virtues promote pro-sustainable actions. In fact, Corral-Verdugo et al. (2015) did not find significant differences in either relationship between SB and virtues that may be classified as "moral" or "intellectual." Therefore, we should look for an alternative answer.

VIRTUES AND UNSUSTAINABILITY

A problem with the schema that conceives virtues as conducive to sustainability is that sometimes character strengths are used for wrongdoing. Older theoretical approaches conceived antisocial and criminal individuals as lacking virtues (see, for instance, Duff, 2013; Rom, 1976); yet a number of studies have found that they exhibit character strengths just as much as other people (Ortiz and Corral-Verdugo, 2017), and that those traits are instrumental for antisocial uses. For instance, creativity is sometimes used for malevolent purposes (Gino and Ariely, 2012), and this form of character strength is governed by the same principles as benevolent creativity (Cropley et al., 2008). Effective leadership by prominent individuals characterizes criminal organizations (Giménez, Ojea, Nieto, Martínez, and Fernández, 2012). Determination and persistence are often characteristics of violent criminals (Athens, 2017); and—no matter how abhorrent and unjustified their behavior is—courage is exhibited by terrorists (Silke, 2004).

In the case of anti-environmental activities, curiosity and love for learning can drive depredatory, exploitative behaviors that harm the natural environment, as in the case of bird hunting (Trinca, Ferrari, and Lees, 2008). Teamwork also is a driver of environmental crimes such as endangered species poaching (Herbig and Griffiths, 2016). Some individuals exhibiting strengths of the humanity virtue act pro-socially but also anti-environmentally (Tal, Hill, Figueredo, Frías, and Corral-Verdugo, 2006). Although spirituality promotes SB, on occasion, religious individuals behave in an anti-ecological way (Hunter and Toney, 2005). This seems to mean that character strengths can be used for good and manifest as "virtuous characters" or for bad and manifest as "vicious characters" (Fowers, 2009).

THE ENVIRONMENT AS MODERATOR OF THE VIRTUES–SUSTAINABILITY RELATIONSHIP

The unspecific result of classifying character strengths either as for "good" or for "bad" begs the question, what determines either outcome? One possible response is the environment.

Contextual factors influence every aspect of an individual's life, including one's actions and behavioral predispositions. A person behaves in certain ways depending on the environment she or he lives in (Barker, 1968). This situation also applies to the physiological functioning of organisms and even to the genetic expression of each individual.

In a recent study, Corral-Verdugo, Ortiz, Frías, Tapia, and Fraijo (2020) tested the idea that environmental unpredictability had a moderating influence on the relationship between character strengths and criminal behaviors. Environmental unpredictability suggests frequent variation in a number of conditions of a setting so that individuals are unable to foresee the future. Environmental unpredictability describes the degree of unanticipated variability in adaptive outcomes of significant behavior. Unpredictable environments interfere with the organism's capacity to solve adaptive problems or escape from those types of aversive events in the future (Winterhalder and Leslie, 2002). Similar negative effects are found in harsh environments predominated by incivility and violence (Brumbach, Figueredo, and Ellis, 2009).

Corral-Verdugo et al. (2020) investigated the self-reported character strengths and SBs in a sample of convicted felons and noncriminal community members. The author found that, in both groups, virtues correlate negatively and significantly with antisocial/delinquent behavior. A test of moderating and mediating effects demonstrated that environmental predictability modified such relationship, decreasing the inhibitory influence of virtues on anti-sociality. The interpretation is that the possession of character strengths is not important to prevent criminality in predictable contexts; however, in unpredictable environments, virtues are necessary to inhibit anti-sociality and delinquent behavior.

Although this is a preliminary study requiring replication, it suggests that the environment can exert a powerful influence on the relationship between virtues and pro-social behavior. Unfortunately, the psychological study of character strengths and virtues has neglected the role environmental factors play in directing those human capacities to either moral or immoral purposes. This also applies to the study of the virtues of sustainability—the set of character strengths that lead individuals to protect their socio-physical environment. While evidence has demonstrated that virtuous people tend to engage in SBs, practically no studies—except for the aforementioned--have been conducted addressing the influence of contextual factors on the promotion or inhibition of pro-sustainability virtues. Since the environment may condition the use of character strengths for good or bad, it is necessary to incorporate the analysis of contextual cues and their effects on the development and utilization of these traits.

CHARACTER STRENGTHS AND VIRTUES AS ADAPTIVE TRAITS

If character strengths are sensitive to environmental pressures, they should serve adaptive purposes. A psychological trait is adaptive if it helps the organism to survive and thrive under specific conditions imposed by the milieu (Bowler, 2003). Making use of character strengths for pro-social and pro-environmental purposes would be adaptive in stable and predictable contexts because individuals may expect future benefits (for them, for others, and for the natural milieu). Delaying immediate gratification, self-controlling impulses, saving resources for the future, and acting altruistically make sense in stable and predictable environments, for example. These are traits of a virtuous person responding to environmental clues (such as stability and predictability), indicating that, ultimately, the environment—as the individual knows it—will persist and she or he can obtain future benefits acting in such a manner. Such environments may thus encourage such virtues as the cooperativeness discussed by Laura Hartman (Chapter 7, this volume). Conversely, in harsher, unstable, and unpredictable environments, the individual may be prompted to act impulsively and egoistically, and without consideration of future consequences because the future is uncertain. Under these conditions, acting to favor one's own interests regardless of the needs of others seems more adaptive than behaving in a pro-social or pro-environmental way (Brumbach et al., 2009). This situation represents one of the foundations for the development of antisocial and criminal behavior. Concluding that criminality emerging from harsh and unpredictable environments is adaptive sounds as controversial as the idea that criminals use their character strengths to commit crimes. However, it makes sense: If adaptation implies the survival and thriving of the individual under specific conditions, it is arguable that taking advantage of others (even with violent methods) is more adaptive than being nice in harsh, unpredictable environments (Brumbach et al., 2009). If this idea proves correct, it would indicate that in harsh, unstable, and unpredictable environments, the adaptive way of life is conducive to the display of "vicious characters," while in more calm, stable, and predictable contexts, the "virtuous characters" (Fowers, 2009) would be more adaptive.

The assumption that the environment influences the relationship between virtues and SB does not negate the impact of additional moderators. Some criminals (especially the white-collar) live in predictable and stable environments; similarly, not all individuals in unstable environments will simply act in their own self-interest or impulsively, so that one could expect their virtues–SB link to be affected by variables other than situational characteristics. Investigating those moderating variables (such as personal history, ideology, attitudes, motives, etc.) is a topic of significant relevance.

THE ENVIRONMENTAL PSYCHOLOGY OF PRO-SUSTAINABILITY VIRTUES

In every society and culture, virtuous characters are preferred over vicious ones, and efforts to promote the former and to discourage the latter are (at least in theory) found in the discourse of authorities, moral leaders, and public policy experts. One manner of achieving this is to generate character strengths in individuals, with the hope that these traits will lead to virtuous behavior and, consequently, to pro-social and pro-environmental actions (i.e., pro-sustainability). Positive psychologists assume that character strengths can be learned, practiced, and cultivated (Park and Peterson, 2008). We already determined the existence of a significant association between virtues and SB (Corral-Verdugo et al., 2015), yet we also concluded that antisocial actions may be associated with character strengths and found (at least one) evidence of a moderating effect of the environment on the relationship between virtues and SB. The turn now is for environmental psychology to address environment–virtuous behavior interactions that could be conducive to sustainability in the form of human well-being and environmental quality.

Since its incipient development, environmental psychology included two clearly distinguishable branches: architectural psychology (AP) and conservation psychology (CP). AP is concerned with the influence that the physical environment exerts on human behavior, while CP mainly focuses on the effect that this behavior has on the integrity of the natural environment (Clayton and Saunders, 2012). The differential approach to environment–behavior interactions of AP and CP sometimes seems to contradict the definition of environmental psychology: the study of the *reciprocal* relations between human behavior and the socio-physical environment (Stokols and Altman, 1987). This definition implies that at any moment the environment influences behavior while, simultaneously, behavior is affecting the environment. Therefore, an integrative approach (Clayton and Saunders, 2012) to the study of the environmental psychology of pro-sustainability virtues would prove useful. The approach would be integrative because it would not address environment–behavior relations in a unidirectional way (either environment → behavior influences or behavior → environment effects). *Transactional theories*, as found in Barker's (1968) and Gibson's (1979) proposals, constitute examples of such an integrative approach. According to those theories, people and settings are inseparable entities, so that the element of analysis should be the person-environment dynamic system in which no unidirectional influences but interdependent relations (i.e., environment ↔ behavior interactions) are sought.

The model of *positive environments* (Corral-Verdugo and Frías, 2016) is another approach that tries to integrate AP and CP, and it may be used to study pro-sustainability virtues. In a positive environment, people see their needs satisfied, receive resources to thrive, and face risks and challenges successfully; they flourish as persons and experience well-being and happiness. Moreover, in this environment, individuals feel compelled to protect the natural and social resources they receive as well as the integrity of their social milieu (including their families, work settings, schools, institutions, cities, etc.). As Corral-Verdugo and Frías establish, "A positive environment provides good things, but also asks for conservation behaviors so that the quality of the present and future environments may be guaranteed" (2016, p. 966). Therefore, positive environments are contexts able to deliver benefits for individuals (and groups) and to instigate behaviors to protect the socio-physical structures on which life depends.

A positive environment contains material and social conditions that enable the satisfaction of human needs; in addition, it produces affordances that instigate self-protective, pro-social, and pro-environmental behaviors. In a positive environment, indicators of socio-physical conservation (or environmental quality), people's well-being and SBs are also detected. Figure 2.1 is a representation

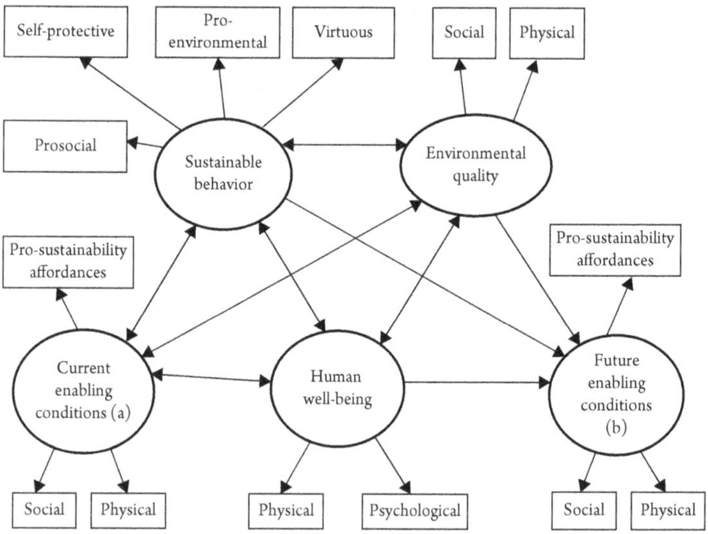

Figure 2.1 A model of environmental positivity including pro-sustainability virtues (adapted from Corral-Verdugo and Frías, 2016).

of the elements of a positive environment and their interrelations. The *current* enabling conditions of environmental positivity ([a] in Figure 2.1) include the present social and physical factors that help to meet human needs as well as affordances for pro-sustainable actions. Human well-being is indicated by, at least, physical and psychological well-being. SB, as previously discussed, encompasses self-protective, pro-social, and pro-environmental activities. Virtuous actions can also be considered a component of SB due to their significant link to pro-social and pro-environmental actions. Social and physical indicators of a well-conserved milieu, in turn, shape environmental quality.

All of these elements are linked to one another through interdependence chains (represented by bidirectional arrows). This means that there are no "original" causes or unidirectional effects of any indicator on other indicators. Enabling socio-physical conditions promote well-being but well-being also influences those enabling conditions (Schulte and Vainio, 2010); SBs affect and are affected by environmental quality (Jenks and Jones, 2010). Human well-being and environmental quality affect each other (Hartig, Kaiser, and Bowler, 2001), and SB and well-being exhibit interdependence (Brown and Kasser, 2005). The unidirectional arrows pointing to future enabling conditions ([b] in Figure 2.1) indicate that current conditions of environmental positivity affect the forthcoming positive conditions, guaranteeing the sustainability of a positive scenario throughout time (Corral-Verdugo and Frías, 2016).

If SBs constantly occur in positive environments, pro-sustainability virtues are also expected to be present in those contexts. In order to manifest as "virtuous characters" (Fowers, 2009), these pro-sustainability virtues and character strengths require that the enabling socio-physical conditions of the environment include signals of predictability, stability, and (relative) calm. One important point here is that pro-sustainability virtues not only manifest through learning and practice; they also require enabling conditions provided by the environment. Some of those conditions include stability, predictability, and (for most of the time) calm; but the social environment must additionally provide civility and pro-sociality models, social ideals of physical and psychological health, and pro-environmental ideals, among other social contexts that facilitate the transformation of character strengths into virtuous traits. Of course, stability, predictability, and calm are occasionally overwhelmed by unexpected negative events in an individual's environment. On occasion, even the most propitious environment experiences risks, adversity, and harshness (as seen in climate calamities, earthquakes, disease, etc.). The positive environment model incorporates those adverse events and situations as challenges that could be successfully faced thanks to specific sets of enabling conditions (social support networks, adequate infrastructure, economic support, etc.). Moreover, within positive environments people use their character strengths to successfully respond to those environmental challenges (Corral-Verdugo and Frías, 2016). This means that politicians, social leaders, researchers, urban designers, religious authorities, policy-makers, and society as a whole should get involved in the design and implementation of those enabling conditions, as well as facilitating the learning, promotion, and cultivation of virtuous traits.

CONCLUDING REMARKS

SB has recently attracted the attention of normative ethics. Three highly influential theoretical approaches—virtue ethics, consequentialism, and deontology—address matters of right action and SB, the latter of which is now considered an important facet of an ethical way of life. Most of the research on the normative ethics of SB centers on deontological and consequentialist aspects of pro-social and pro-environmental actions, to the detriment of studying the virtuous aspect of pro-sustainability orientation. One possible reason for this lack of research is that, until recently, SB was not considered of paramount importance within the set of character strengths and virtues exhibited by people.

In ancient times, being virtuous was typically focused on defending honor, one's own people, and culture. Strangers were not necessarily seen as fellow persons who deserved consideration, sympathy, and help: "Love thy neighbor" was literal; the fellow was genetically, spatially, or culturally close, and the "others" were seen as alien. Therefore, altruism, as we currently conceive it, was mostly limited to loving and assisting close people (something that, unfortunately, in several modern communities continues; see Abascal and Baldassarri, 2015). The natural environment was—at least in the Western tradition—much stranger than non-fellow persons were. In many cultures, plants and animals were (and are) seen as "ordered to man's use," as Thomas Aquina argued in his *Summa Contra Gentiles* treatise (for a very different perspective, see Pankaj Jain's discussion of Jain ethics, Chapter 5, this volume). The Bible's book of Genesis (1: 27–28) states: "God created man in his own image . . . and God said unto them, be fruitful, and multiply, and replenish the earth, and subdue it." Even Aristotle (*Politics,* Book 1, chapter 8) assured that "nature has made all things specifically for the sake of man." One cannot expect a sustainability virtue to emerge from those sets of beliefs and social conditions.

In modern times, because of the dramatic changes in Earth's biosphere, philosophy, religion, and morality coincide in considering stranger humans, animals, plants, and even the inorganic substrate of Earth as deserving protection. The aim of morality and virtuosity has also changed in correspondence with this cultural shift, at least in some modern cultural institutions. Pope Francis (2015), for instance, issued his first encyclical (*Laudato Si, On Care for our Common Home*) and devoted it to the moral imperative of protecting our planet.

Consequently, character strengths and virtues are currently considered traits that may serve the purpose of sustainable development. Promoting the conditions that facilitate the pro-social and pro-environmental use of character strengths and virtues should be an imperative in every community and culture throughout the world.

Environmental psychology may help in this endeavor, by studying environment-behavior transactions that result in the development and use of the virtuous characters that can be labeled "pro-sustainability virtues." These traits would be identified as manifestations of excellence or character resulting in personal flourishing, pro-social acting, and environmentally protective behavior.

Pro-sustainability virtues may be produced in positive environments, whereas vicious characters are expected to predominate in negative human scenarios (Corral and Frías, 2016; Ortiz, 2017). We already know how character strengths and virtues can be cultivated and developed (Park and Peterson, 2008); now we need to generate and maintain the conditions that transform those traits into virtuous (i.e., pro-sustainability) characters. Virtuosity was sometimes considered to have an intrinsic value, implying that the virtuous elevation of a person is worth it per se. Sumner (1998, p. 22), for instance, states, "The perfectionist value of a life is conceptually independent of how well the life is going for its owner." This does not mean that virtuosity cannot provide a source of well-being for oneself—in addition to providing good results for others and the environment. It just means that other sources of well-being exist aside a virtuous life and being virtuous is worthwhile by itself. Since SB implies acting for a moral reason, the deontological purpose of being a rightful individual would reinforce even more the desire to protect one's self, others, and the physical environment. This conclusion is supported by Devettere (2002, p. 13), who states, "Ethics originates in our natural desires for doing well and living well." Following this logic, there is no contradiction between acting rightful, being pro-sustainable, and living well; on the contrary, these three facets seem to act synergistically. Therefore, the most important challenge is to set the conditions that enable the use of character strengths for pro-sustainability purposes, and they are to be found, fundamentally, in the environment and in the interaction between those conditions and virtuous behavior.

Scientific research plays an important role in elucidating those conditions and how they can be designed and implemented. Although environmental psychology has a great deal to do in this regard, it is clear that interdisciplinary studies should be conducted to investigate the cultural, social, moral, ethical, ecological, legal, and psychological (among many other) aspects involved in the emergence and sustenance of pro-sustainability virtues.

REFERENCES

Abascal, M., and D. Baldassarri. 2015. "Love thy neighbor? Ethnoracial diversity and trust reexamined." *American Journal of Sociology, 121*, 722–782.

Athens, L.H. 2017. *The Creation of Dangerous Violent Criminals* (2nd ed.). New York: Routledge.

Baehr, J. 2011. *The Inquiring Mind: On Intellectual Virtues and Virtue Epistemology*. Oxford, UK: Oxford University Press.

Ballantyne, R., J. Fien, and J. Packer. 2001. "Program effectiveness in facilitating intergenerational influence in environmental education: Lessons from the field." *Journal of Environmental Education, 32*, 8–15.

Barker, R. 1968. *Ecological Psychology: Concepts and Methods for Studying the Environment of Human Behavior.* Palo Alto, CA: Stanford University Press.

Baum, C., and C. Gross. 2017. "Sustainability policy as if people mattered: developing a framework for environmentally significant behavioral change." *Journal of Bioeconomics, 19*, 53–95.

Bermúdez, J. 1999. "Personality and health-protective behavior." *European Journal of Personality, 13*, 83–103.

Bowler, P.J. 2003. *Evolution: The History of an Idea* (3rd ed.). Berkeley: University of California Press.

Brown, K.W., and T. Kasser. 2005. "Are psychological and ecological well-being compatible? The role of values, mindfulness, and lifestyle." *Social Indicators Research, 74*, 349–368.

Brumbach, B., A. Figueredo, and B. Ellis. 2009. "Effects of harsh and unpredictable environments in adolescence on development of life history strategies: A longitudinal test of an evolutionary model." *Human Nature, 20,* 25–51.

Clayton, S., and C. Saunders. 2012. "Introduction: Environmental and conservation psychology." In *Oxford Handbook of Environmental and Conservation Psychology,* edited by S. Clayton, 1–6. New York: Oxford University Press.

Clifton, D. 2012. "Sustainable business: Are we heading in the right direction?" *Sustainability, 4,* 586–603.

Corral-Verdugo, V. 2012. "The positive psychology of sustainability." *Environment, Development and Sustainability, 14,* 651–666.

Corral-Verdugo, V., J. Caso, and M. Frías. 2017. *Psicología del Cambio Climático* [Psychology of climate change]. Mexico City: Pearson.

Corral-Verdugo, V., and M. Frías. 2016. "The sustainability of positive environments." *Environment, Development and Sustainability, 18,* 965–984.

Corral-Verdugo, V, M. Frías, and C.H. Garcia-Cadena. 2010. "Introduction to the psychological dimensions of sustainability." In *Psychological Approaches to Sustainability,* edited by V. Corral-Verdugo, C. Garcia, and M. Frias, 3–18. New York: Nova Science Publishers.

Corral-Verdugo, V., C. García, L. Castro, I. Viramontes, and R. Limones. 2010. "Equity and sustainable lifestyles." In *Psychological Approaches to Sustainability,* edited by V. Corral-Verdugo, C. Garcia, and M. Frias, 185–204. New York: Nova Science Publishers.

Corral-Verdugo, V., D. González, M. Rascón, and V. Corral-Frías. 2016. "Intrinsic motives of autonomy, self-efficacy, and satisfaction associated with two instances of sustainable behavior: frugality and equity." *Psychology, 7,* 662–671.

Corral-Verdugo, V., S. Hess, B. Hernández, and E. Suárez. 2002. "Los fundamentosy la estructura de la acción proecológica, medidos en una escala de conductas protectoras del ambiente" [Foundations and structure of pro-ecological actions, assessed with a scale of environmentally protective behaviors]. In *Conductas Protectoras del Ambiente* [Environmentally protective behaviors], edited by V. Corral-Verdugo, 71–88. Mexico City: Plaza and Valdés.

Corral-Verdugo, V., A. Ortiz, M. Frías, C. Tapia, and B. Fraijo. 2020. "Environmental predictability influences the relationship between virtues and antisocial/delinquent behavior." *Psychology, Crime and Law, 26,* 148–166.

Corral-Verdugo, V., C. Tapia., and A. Ortiz. 2015. "On the relationship between character strengths and sustainable behavior." *Environment and Behavior, 47,* 877–901.

Cropley, D., J. Kaufman, and A. Cropley. 2008. "Malevolent creativity: A functional model of creativity in terrorism and crime." *Creativity Research Journal, 20,* 105–115.

Devettere, R. 2002. *Introduction to Virtue Ethics.* Washington, DC: Georgetown University Press.

Dickerson, C., R. Thibodeau, E. Aronson, and D. Miller. 1992. "Using cognitive dissonance to encourage water conservation." *Journal of Applied Social Psychology, 22,* 841–854.

Duff, R. 2013. "Virtue, vice, and the criminal law—A response to Huigens and Yankah." In *Law, Virtue, and Justice,* edited by A. Amaya, and H. Lai, 195–214. Portland, OR: Hart.

Fowers, B.J. 2005. *Virtue and Psychology: Pursuing Excellence in Ordinary Practices.* Washington, DC: APA Press.

Fowers, B.J. 2009. "Virtue ethics." In *The Encyclopedia of Positive Psychology,* edited by S.J. López, 1011–1016. Chichester, UK: Wiley-Blackwell.

Fraijo, B., V. Corral-Verdugo, C. Tapia, and D. González. 2010. "Promoting proenvironmental competency." In *Psychological Approaches to Sustainability,* edited by V. Corral-Verdugo, C. Garcia, and M. Frias, 225–246. New York: Nova Science Publishers.

Gibson, J. 1979. *The Ecological Approach to Visual Perception.* Boston: Houghton Mifflin.

Giménez, A., H. Ojea, L. Nieto, J. Martínez, and S. Fernández. 2012. "Estructuras y liderazgo en cuatro redes españolas dedicadas al tráfico de drogas" [Structures and leadership in four drug-trafficking Spanish networks]. *Revista Española de Investigación Criminológica, 10,* 1–32.

Gino, F., and D. Ariely. 2012. "The dark side of creativity: Original thinkers can be more dishonest." *Journal of Personality and Social Psychology, 102,* 445–459.

Gouveia, V. 2002. "Self, culture, and sustainable development." In *Psychology of Sustainable Development,* edited by P. Schmuck and P.W. Schultz, 151–174. Norwell, MA: Kluwer.

Grimm, S. 2009. "Epistemic normativity." In *Epistemic Value,* edited by A. Haddock, A. Millar, and D. Pritchard, 243–264. Oxford, UK: Oxford University Press.

Han, H. 2015. "Virtue ethics, positive psychology, and a new model of science and engineering ethics education." *Science and Engineering Ethics, 21,* 441–460.

Hargreaves, T. 2010. "Putting Foucault to work on the environment: Exploring pro-environmental behavior change as a form of discipline." CSERGE working paper. EDM, No. 10-11, http://hdl.handle.net/10419/48799.

Hartig, T., F. Kaiser, and P. Bowler. 2001. "Psychological restoration in nature as a positive motivation for ecological behavior." *Environment and Behavior, 33,* 590–607.

Herbig, F., and M. Griffiths. 2016. "Conservation crime and rhinoceros poaching: from ancient custom to modern dilemma." *Acta Criminologica: Southern African Journal of Criminology, 29,* 129–142.

Hilbig, B.E., I. Zettler, M. Moshagen, and T. Heydasch. 2013. "Tracing the path from personality—via cooperativeness—to conservation." *European Journal of Personality, 27,* 319–327.

Hirsh, J.B. 2010. "Personality and environmental concern." *Journal of Environmental Psychology, 30,* 245–248.

Hsu, S.J. 2004. "The effects of an environmental education program on responsible environmental behavior and associated environmental literacy variables in Taiwanese college students." *Journal of Environmental Education, 35,* 37–48.

Hunter, L., and M. Toney. 2005. "Religion and attitudes toward the environment: A comparison of Mormons and the general U.S. population." *The Social Science Journal, 42,* 25–38.

Hursthouse, R., and G. Pettigrove. 2016. "Virtue ethics." In *The Stanford Encyclopedia of Philosophy* (Winter 2016 edition), edited by E.N. Zalta. https://plato.stanford.edu/archives/win2016/entries/ethics-virtue/.

Iwata, O. 2002. "Coping style and three psychological measures associated with environmentally responsible behavior." *Social Behavior and Personality, 30,* 661–669.

Jenks, M., and C. Jones, eds. 2010. *Dimensions of the Sustainable City.* New York: Springer.

Jordan, K., and K. Kristjánsson. 2015. "Sustainability, virtue ethics, and the virtue of harmony with nature." *Environmental Education Research, 23,* 1205–1229.

Kaiser, F., P.W. Schultz, J. Berenguer, V. Corral-Verdugo, and G. Tankha. 2008. "Extending planned environmentalism: Anticipated guilt and embarrassment across cultures." *European Psychologist, 13,* 288–297.

Kals, E., D. Schumacher, and I. Montada. 1999. "Emotional affinity toward nature as a motivational basis to protect nature." *Environment and Behavior, 31,* 178–202.

Kawall, J. 2009. "In defense of the primacy of virtues." *Journal of Ethics and Social Philosophy,* 3, 1–21.

Lourdel, N., J. Martin, and O. Bérerd. 2006. *Overcoming obstacles to understanding sustainable development—An approach based on personal experiences.* EESD 2006 Conference. Lyon, France.

Mahoney, A., R. Carels, K. Pargament, A. Wachholtz, L. Edwards, M. Kaplar, and R. Frutchey. 2005. "The sanctification of the body and behavioral health patterns of college students." *The International Journal for the Psychology of Religion, 15,* 221–238.

Malott, R.W. 2010. "I'll save the world from global warming—Tomorrow: Using procrastination management to combat global warming." *The Behavior Analyst,* 33, 179–180.

Milfont, T.L., J. Wilson, and P.K.C. Diniz. 2012. "Time perspective and environmental engagement: A meta-analysis." *International Journal of Psychology, 46,* 209–218.

Norton, B.G., and A. Thompson. 2014. "Ethics and sustainable development: the virtues of an adaptive approach to environmental choice." In *Handbook of Sustainable Development,* edited by G. Atkinson, S. Dietz, E. Neumayer and M. Agarwala, 105–124. London: Edward Elgar.

Ortiz, A. 2017. *Dimensiones ambientales y fortalezas de carácter como auspiciadoras de la conducta anti y pro social [Environmental dimensions and character strengths as promoters of anti and pro social behavior]*. Doctoral Dissertation. University of Sonora, Mexico.

Ortiz, A., and V. Corral-Verdugo. 2017. "Virtudes y fortalezas de carácter en personas recluidas en un sistema penitenciario" [Virtues and character strengths in persons reclused in a penitentiary system]. *Psicumex, 7*, 80–92.

Park, N., and C. Peterson. 2008. "The cultivation of character strengths." In *Teaching for Wisdom: Cross-Cultural Perspectives on Fostering Wisdom,* edited by M. Ferrari and G. Potworoski, 59–77. New York: Springer Science.

Pato, C., and V. Corral-Verdugo. 2013. *Cuidado consigo mesmo e cuidado delambiente [Oneself protection and environmental protection]*. Paper presented at the XXXIV Interamerican Congress of Psychology. Brasilia, Brazil.

Pelletier, L.G. 2002. "A motivational analysis of self-determination for pro-environmental behaviors." In *Handbook of Self Determination Research,* edited by E.L. Deci and R.M. Ryan, 205–232. Rochester, NY: University of Rochester Press.

Peterson, C., and M.E.P. Seligman. 2004. *Character Strengths and Virtues: A Handbook and Classification*. Washington, DC: American Psychological Association.

Pope Francis. 2015. *Laudato Si, On Care for our Common Home.* Retrieved from http://w2.vatican.va/content/dam/francesco/pdf/encyclicals/documents/papa-francesco_20150524_enciclica-laudato-si_en.pdf.

Rezvani, Z., J. Jansson, and M. Bengtsson. 2017. "Cause I'll feel good! An investigation into the effects of anticipated emotions and personal moral norms on consumer pro-environmental behavior." *Journal of Promotion Management, 23*, 163–183.

Rom, P. 1976. "Criminals are discouraged persons." *Self and Society, 4*, 22–23.

Schulte, P., and H. Vainio. 2010. "Well-being at work: Overview and perspective." *Scandinavian Journal of Work, Environment and Health, 36*, 422–429.

Seligman, M., and M. Csikszentmihalyi. 2000. "Positive psychology. An introduction." *American Psychologist, 55*, 5–14.

Sevillano, V., J.I. Aragonés, and P.W. Schultz. 2010. "Altruism and beyond: The motivational bases of pro-environmental behavior." In *Psychological Approaches to Sustainability,* edited by V. Corral-Verdugo, C. Garcia and M. Frias, 161–184. New York: Nova Science Publishers.

Silke, A. 2004. "Courage in dark places: Reflections on terrorist psychology." *Social Research, 71*, 177–198.

Stern, P.C. 2000. "Toward a coherent theory of environmentally significant behavior." *Journal of Social Issues, 56*, 407–424.

Stokols, D., and I. Altman, eds. 1987. *Handbook of Environmental Psychology.* New York: Wiley.

Sumner, L.W. 1998. "Is virtue its own reward?" In *Virtue and Vice,* edited by E. Paul, F. Miller, and J. Paul, 18–36. Cambridge, UK: Cambridge University Press.

Tal, I., D. Hill, A.J. Figueredo, M. Fría, and V. Corral-Verdugo. 2006. "An evolutionary approach to explaining water conservation behavior." *International Journal of Environmental Psychology, 7,* 7–27.

Tapia, C., V. Corral-Verdugo, B. Fraijo, and F. Durón. 2013. "Assessing sustainable behavior and its correlates: a measure of pro-ecological, frugal, altruistic and equitable actions." *Sustainability, 5,* 711–723.

Thompson, P.B. 2007. "Agricultural sustainability: What it is and what it is not." *International Journal of Agricultural Sustainability, 5,* 5–16.

Trinca, C., S. Ferrari, and A. Lees. 2008. "Curiosity killed the bird: arbitrary hunting of Harpy Eagles *Harpia harpyja* on an agricultural frontier in southern Brazilian Amazonia." *Cotinga, 30,* 12–15.

Vaillant, G.E. 2008. "Positive emotions, spirituality and the practice of psychiatry." *Mens Sana Monographs, 6,* 48–62.

Van Lange, P. 2000. "Cooperation and competition." In *Encyclopedia of Psychology, Vol. 2,* edited by A.E. Kazdin, 296–300. Washington, DC, New York: American Psychological Association and Oxford University Press.

Vucetich, J., and M. Nelson. 2010. "Sustainability: virtuous or vulgar?" *Bioscience, 60,* 539–544.

Werder, O. 2006. "Influences on the recycling behavior of young adults: Avenues for social marketing campaigns." In *The Environmental Communication Yearbook,* Vol. 3, edited by P. Depoe, 77–96. Mahwah, NJ: Lawrence Erlbaum.

Winterhalder, B., and P. Leslie. 2002. "Risk-sensitive fertility: the variance compensation hypothesis." *Evolution and Human Behavior, 23,* 59–82.

Chapter 3

Aristotelian Virtue Education and Education for Sustainable Development

Prospects and Problems

MATT FERKANY

INTRODUCTION

The sustainability movement has viewed education, including formal education in schools and universities, as a key component of social change from its inception. Changes in citizens' values, attitudes, and behavior, not just knowledge and understanding, have also been central goals of the movement's major educational declarations (e.g., Chapter 36 of Agenda 21; the U.N. Decade of Education for Sustainable Development). For all that, ideals and models of sustainability education rarely mention the virtues (for an exception, see Curren and Metzger, 2017).

This seems odd. Understood as Aristotelian <u>*hexis*</u>—relatively stable conditions or states—virtues might be useful ways to capture

Matt Ferkany, *Aristotelian Virtue Education and Education for Sustainable Development* In: *The Virtues of Sustainability*. Edited by: Jason Kawall, Oxford University Press (2021). © Oxford University Press.
DOI: 10.1093/oso/9780190919818.003.0003

whole suites of sustainability-related dispositions, values, attitudes, and behaviors. Yet little scholarship addresses itself to what it would mean for the shape and content of sustainability education in schools and universities if virtue were treated as an aim.

This chapter seeks to make a start on closing this gap in the literature. Taking widely shared ideas in Aristotelian virtue ethics as my starting point, I discuss how Aristotelian virtue education (AVE) can fit into and improve the mission of education for sustainable development (ESD). Reading the goals of ESD through AVE helps to organize those goals and improves our understanding of how they are to be achieved. It highlights the significance of human flourishing as the ultimate goal—and possibly the measure—of sustainability; the moral dimension of sustainable development; and the importance of experiencing the best things in life and of practicing sustainability virtue in a virtue-loving social environment.

I then address two potential criticisms of this approach. I conclude with some observations concerning the near-term potential of current institutions in the United States to advance the cause of sustainable development given the demands of AVE. If the story told in this chapter is plausible, schools and universities have a long way to go before they can be expected to impart the sustainability virtues to citizens.

EDUCATION FOR SUSTAINABLE DEVELOPMENT

The sentiment that education, including that intended to influence people's beliefs, values, and attitudes, is crucial to sustainable change might be as old as environmental movements themselves. It is present in the work of early conservationists, such as Pinchot, who identified educating the public as the key to sparking the conservation movement in *The Fight for Conservation* (1910). Rachel Carson wrote entire books for the purpose of guiding environmental education (1965). The ultimate aim of Aldo Leopold's land ethic was a cultural revolution in which humans come to see themselves as plain members and citizens of the land alongside the rest of the biotic community (1949).

From its earliest days the sustainable development movement followed suit, promulgating an educational ideal that has come to be known as *education for sustainable development* (ESD). In early articulations of this ideal, the core value is development that meets the needs of the present without compromising the ability of future generations to meet their own needs. The role of education, "Both formal and non-formal," is "changing people's attitudes so that they have the capacity to assess and address their sustainable development concerns" and "[achieve] environmental and ethical awareness, values and attitudes, skills and behavior consistent with sustainable development and for effective public participation in decision-making" (UNCED, n.d., chap. 36).

More recently the UNESCO *Decade of Education for Sustainable Development* (DESD) defines sustainable development as "the will to improve everyone's quality of life, including that of future generations, by reconciling economic growth, social development and environmental protection" (UNESCO, 2005, para. 5). The "overall goal of the initiative [was] to integrate the values inherent in sustainable development into all aspects of learning to encourage changes in behavior that allow for a more sustainable and just society for all" (UNESCO, 2006, para. 1). This goal was to be achieved through a transformation of educational institutions and practices so that they do (among other things) the following:

1. *Teach respect*: Teach respect for "human dignity and human rights of all people; for the human rights of future generations

and intergenerational responsibility; respect and care for the greater community of life in all its diversity which involves the protection and restoration of Earth's ecosystems; respect for cultural diversity and a commitment to build locally and globally a culture of tolerance, non-violence and peace." (UNESCO, 2006, p. 16)

2. *Teach complexity*: Teach "the complexity and interconnectedness of problems such as poverty, wasteful consumption, environmental degradation, urban decay, population growth, health, conflict and the violation of human rights that threaten our future" (UNESCO, 2002, p. 10).
3. *Teach democracy and peace*: Teach in a manner that is participatory in both pedagogical method—such that "learners participate in decisions on how they are to learn" (UNESCO, 2006, p. 17)—and in purpose, so that learners can "be committed to promoting democracy in a society without exclusion and where peace prevails" (UNESCO, 2005, sec. 3).
4. *Teach critical thinking*: Teach critical thinking and problem solving that are "values-driven" so that the values and principles of sustainable development "are made explicit [and] can be examined, debated, tested and applied" (UNESCO, 2006, p. 17).
5. *Infuse ESD everywhere*: Do all this in an interdisciplinary and holistic way, so that "learning for sustainable development [is] embedded in the whole curriculum, not as a separate subject" at all levels. (UNESCO, 2006, p. 17).

The *Decade* initiative has ended only recently, and its impact on global educational practice remains unclear, though it appears to have had little impact in North America (Feinstein and Kirchgasler, 2015; Nolet, 2009). Within academia, the *Decade* had the salutary effect of sparking lively debate about the aims, purposes, and methods of sustainability education. The debate yielded many variants of ESD, some taking the model to task for papering over the tensions between the values of economic growth, environmental protection, and social justice (Curren and Metzger, 2017; Schlottmann, 2008); for underappreciating the inherent worth of species and the interconnectedness of life on Earth (Stone and Barlow, 2005); and for lacking a social theory linking the domination of some humans by others to their proclivity for dominating nature (Martusewicz, Edmundson, and Lupinacci, 2011).

These criticisms raise serious substantive issues, some of which will be addressed herein. In general, however, the many variations of ESD share the transformative goal of shaping students' sustainability-related attitudes, values, and behaviors through some variant of the commitments 1–5 earlier. To that extent, ESD is a fair proxy for evaluating the advantages of virtues education for sustainability education. I believe there are several. First, however, a word about virtue.

VIRTUE AND SUSTAINABILITY

Virtues are character traits belonging to morally and intellectually good persons. Like personality traits, virtues are thought to be psychologically complex, involving whole suites of dispositions to want, think, act, and feel courageously, justly, generously, and so on. For Aristotelians, however, virtues are unlike personality traits in that they involve skill-like exercise of the will toward the end of flourishing or living well as a human being. Virtuous persons do not simply do the right thing by accident, begrudgingly, or because they will get something out of it. They act from virtue, that is, a stable understanding of the good of virtues and from love of that good for its own sake. On some views, they must also know why virtue demands what it does, in the sense that they could, if asked, say why (Curzer, 2012; Hills, 2015).

Aristotelians ordinarily delineate the virtues by reference to what it takes for humans to flourish (Hursthouse, 1999; Watson, 1990). This should not be read as stating that virtues are mere instruments or means of achieving some good theoretically achievable independently of virtue. For Aristotelians, a good life does involve certain *basic external goods*, those external to virtue, such as health, a safe environment, and basic wealth. For this reason Aristotle denies that "one calls a man happy who meets with misfortunes like Priam's, and comes to a miserable end" (*Nichomachean Ethics*, 1.9). But the role of virtue in well-being is not external in this way; living well consists in being a good person barring "misfortunes like Priam's" (and presumably lesser but equally out-of-our-control misfortunes as those befalling Trojan emperors).

Out of this basic framework somewhat different ideals of virtue can be developed. The precise ideal one develops will depend upon one's view of human nature, of well-being, and of the obstacles to living well people face given the conditions of their lives. Drawing on commonsense observation and empirical research, Aristotelians commonly maintain that the best things in life include those that exercise people's powers as rational and creative social animals, such as (but not limited to) virtue itself, social bonds (i.e., friendship and family ties), autonomy, physical health and fitness, and delight in stimulating intellectual and creative activity (Kraut, 2007; Nussbaum, 2000). Many believe that attaining these goods is impossible without background social and environmental conditions protecting the basic external goods for all people, and for this reason call for some form of democratic egalitarian justice as a condition for the pursuit of virtue (Curren, 2000; Nussbaum, 2000). But they also see the relationship between background conditions and the virtues as reciprocal (call this the *reciprocity thesis*): Virtue is impossibly difficult without basic social justice, but justice is impossibly difficult to enjoy when people are crudely egoistic, disloyal, dishonest, stingy, intemperate, cowardly, morose, lazy, and so on. For this reason, Aristotelians often part ways with liberal orthodoxy in viewing the shaping of people and their character—toward the end of virtue—as a legitimate enterprise for the state (Curren, 2000; Gutmann, 1999).

However exactly sustainability is conceptualized, it at least concerns human good and the requirements of protecting or promoting that good indefinitely given factors such as global scarcity of resources, limits of regional and global systems to support development and absorb waste, diversity of peoples and ways of life, global North/South inequality, and the history of colonialism behind it. Complicating problems of sustainability is the absence of a state-like authority having power to compel nations to cooperate and the complexity of the causes of global unsustainability, as well as the complexity—and frequently very technical nature—of the information needed to make sustainability-related decisions.

Dealing with problems of sustainability will require well more than virtue in the citizens of the world. Leaders and citizens of individual nations will need to make sustainability a policy priority. Nations will then have to find ways to set fair terms for global cooperation in ways that ramify back into sustainability policy at the level of their own national policies. However, it is very unlikely that such an arrangement will come to be if citizens of democratic nations are unwilling to permit national policy to be set partly in recognition of the fair terms of global cooperation (Curren and Metzger, 2017). If the reciprocity thesis is correct, virtue may have an outsized role to play in the achievement (or not) of global sustainability in the absence of a global coercive power to enforce cooperation.

A full discussion of the sustainability virtues suggested by Aristotelianism is beyond the scope of this chapter. Different sustainability challenges may require reconceptualizing traditional virtues, like justice, in novel ways or call upon qualities having practical relevance that is previously unimagined, such as creativity.[1] However, it will aid discussion (of the merits of reconceptualizing ESD in terms of AVE) to specify, however roughly, a few particular (sets of) virtues. I focus on a few that, from an Aristotelian perspective, global citizens will likely need in confronting sustainability problems. These include (a) virtues of cosmopolitan justice; (b) virtues of right attachment to people, places, and things; and (c) virtues of proper testimonial credulity.[2]

Virtues of cosmopolitan justice include fairness, cooperativeness, goodwill, responsibility, and limited patriotism. Achieving fair terms of global cooperation in the resolution of sustainability problems will frequently require substantial compromise of perceived individual or group self-interest.[3] Citizens of the global North in particular, as heirs to colonialism and associated development, must be prepared to accept a larger share of the responsibility for curbing unsustainability while facilitating development of the global South. Patriotic attachment to the places of home—and the sense of citizen identity and self that go with this—can be a powerful motive to support conservation, but it must not take the form of a national sense of uniqueness and superiority.

Virtues of right attachment to people, places, and things include virtues of love, temperance (or moderation), frugality, and (again) limited patriotism. Virtues of right attachment are needed if people are to be motivated to care about the long-term good of home as well

1. On the practical relevance of creativity, see Wright, Chapter 10, this volume.
2. Important others suggested in other chapters of this book also fit the Aristotelian model. See again chapters by Sarah Wright (Chapter 10) on creativity and open-mindedness (on my list); Jason Kawall's chapter on patience (Chapter 8); and Laura Hartman's chapter on cooperativeness (Chapter 7; similar to my virtues of cosmopolitan justice).
3. For a somewhat different take on the virtues relevant here, see Hartman, Chapter 7, this volume.

as the good of people in distant places and times, and also to avoid materialist habits of consumption that are known to be bad both for the planet and for our own well-being (Gambrel and Cafaro, 2010; Kasser, 2002).

Virtues of proper testimonial credulity include open-mindedness and distinctly intellectual forms of justice, autonomy, humility, and charity.[4] Sustainability decision making will require citizens to sort through competing accounts of scientific information in a highly politicized media environment, one where powerful interests seek to press their advantage through coordinated campaigns of misinformation (such as in the case of climate change). Without these virtues, citizens of democratic societies are unlikely to properly identify credible sources of news or scientific information or, with credible information in hand, choose well in light of it.

ARISTOTELIAN VIRTUE EDUCATION

Virtues ordinarily have to be learned, refined, and maintained through thoughtful practice and a lifelong drive to improve as persons (Annas, 2011). Aristotle maintained that becoming virtuous requires habitually conforming one's behavior to virtue, preferably under the guidance of a (more) virtuous person (*Nichomachean Ethics*, 2.1). Teachers of virtue model virtue for the learner and use praise or blame to motivate virtue-conforming behavior and the internalization of standards of virtue. Teaching or direct instruction in the

4. For a similar account, see Roberts and Wood (2007). The intellectual virtues more generally (such as love of knowledge, rigor) are arguably important sustainability virtues given the complexity of sustainability challenges. I focus later on the virtues on proper testimonial credulity in part to have a somewhat more specific target for later discussion, and in part because of the foundational role of these virtues in enabling successful beginnings to intellectual inquiry.

virtues, that is, reasoned discussion of virtue and the goods of virtue, must come later. Without practice and the experience practice provides, the learner's desires are not aligned with virtue, and she lacks first-hand insight into the good of virtue; thus, she cannot be motivated to conform her behavior to virtue through reasoning alone.

Following Aristotle, modern character educators generally understand becoming virtuous to involve a lifelong process of internalizing norms for virtue—as demonstrated by models or the study of models—through practice, and later through direct instruction, in a supportive social environment and virtue-loving culture. Altogether the development of virtue requires practice, experience, and direct, explicit study of virtue concepts and ideas working together, all in a virtue-loving social environment (Kristjánsson, 2015).

AVE entails several principles for curriculum, pedagogy, and institutional design. Schools and teachers following this approach will model sustainability virtues (so far as possible) and (in later years especially) engage students in some direct reflection on the virtues. Modeling involves schools and teachers being and doing what they want students to be or become. Direct reflection on the virtues involves naming them, thinking about what they are or the circumstances for their exercise, and about the values they serve, typically through the study of exemplars (such as Rachel Carson or Aldo Leopold, perhaps). They also will confront students with opportunities to practice virtues. Successful schools will be academically challenging, requiring students to stretch themselves and exercise virtues of curiosity, open-mindedness, and intellectual carefulness, humility, autonomy, and the like.[5] In the service of the character virtues, the curriculum will be expansive and might include as requirements "extracurriculars" such as music, art, sport, community service,

5. For a model of a school having intellectual virtue as an organizing mission, see the website of the Intellectual Virtues Academy in Long Beach, California (IVA Administration, n.d.).

engagement with the broader community outside the school, and forms of outdoor or place-based experiential learning. Children are unlikely to learn the superior value of more sustainable activities (e.g., creativity or community involvement) to shopping (say) if the curriculum is reduced to passing standardized tests in reading and math. This curriculum will also be delivered in an engaging way so that students learn in an active, student-centered way; it will be approached with the purpose of imparting love of virtue for its own sake, independent critical thinking skills, and a "critical spirit," not mere absorption of knowledge. To do so, schools must to the extent possible become communities of virtue that filter out and seek to counteract the forces outside of school supporting vice. School leadership must demonstrate the virtues in their own talk and behavior, and the overall culture of the school should encourage virtue and make it rewarding while discouraging and imposing costs on vice.[6]

Much of what is required by this ideal is commonsensical and consistent with long-standing views in educational theory since the progressivism of Maria Montessori and John Dewey. Educators and schools will, so far as possible, model and do what they wish the learner to become and do, and they will put the learner in position to become by way of active doing. But some of it is more radical. Requiring extracurriculars, for instance, would call for major reform of public school funding structures in the United States, which at present are woefully inadequate even to selective participation. Because virtue learning requires a virtue-loving environment, the educational environment also extends well beyond school/university boundaries to include the broader culture. Aristotelian virtue educators are thus

6. However, the appropriate ways of making virtue rewarding and vice costly are controversial. It is particularly controversial whether the use of external incentives (i.e., good or bad consequences beyond those that directly result from the exercise of virtue/vice) is appropriate or effective. For a critique of incentives, see Kohn (1997). For a limited defense in some contexts, see Ferkany and Creed (2014).

committed to the reform of that broader culture along whatever lines are needed, for example toward greater economic equality,[7] where opportunity to find meaningful work and enjoy the best things in life is grossly unequal outside school walls.

ADVANTAGES OF ARISTOTELIAN VIRTUE EDUCATION FOR ESD

The principles of AVE suggest several general parameters for ESD at the institutional (school/university) level, some consistent with existing approaches, some pressing in newer directions. Advocates of ESD agree that schools must go beyond teaching *about* sustainability ideas and principles and involve students, in some way or another, in practices of sustainability such as school waste management, food sourcing, or improving energy use.[8] And they agree that this should be done across all grade levels and integrated across the curriculum, not taught only as a separate subject.

AVE adds that institutional politics, teachers, and major school functions should model sustainable practices as well. Institutional politics must ensure that access to and participation in quality education are justly distributed; a citizenry cannot take on the values of sustainability if only a portion of its populace ever encounters them in any meaningful way. Teachers must be seen by students to value sustainability through concrete behaviors, such as how they manage the classroom or get themselves to work. Practices and functions that

7. Aristotle in *The Politics* is particularly clear about the importance of rough economic equality to virtue among citizens of stable democracies. See, for example, *Politics* II:vii; IV:xi; or VI:v. But see also the contemporary reading of Aristotle in Curren (2000).

8. For different views that nevertheless share the goal of transforming students' sustainability behavior (and so go beyond teaching about sustainability), compare the views in Curren and Metzger (2017) to Orr (1991).

are wasteful, undemocratic, or harmful to community and individual well-being—the vast inequities in school funding in the United States, for example—should be abolished or reformed.[9] Finally the ethical significance of these practices—their link to living well and specific virtues—must be made more and more explicit and discussed critically as learners mature.

Approaching ESD through this basic framework has several benefits. I discuss four in this section.

AVE Brings More Coherence to the Values of ESD

A common criticism of ESD is that its ecumenical conception of sustainability is confused. Thus its deployment as an educational ideal is liable to lead to inconsistent policy, classroom confusion, and overall weak sustainability messaging. One merit of Aristotelianism is that it helps settle this controversy on one side in a certain way.

The "three-legged stool" conception of sustainability might be seen as an effort to settle the controversy between "strong" and "weak" ideals of sustainability. Weak and strong ideals each involve different answers to the questions "What is to be sustained?" and "How is sustainable activity to be measured?" According to a weak conception, no natural resources in particular need to be sustained, including other species, provided that each generation passes on sufficient economic and technological wherewithal for the next generation to meet its own needs, where needs are typically measured in terms of GDP. Natural resources on this view are fungible with respect to human interests or well-being, and in this sense weak sustainability is neutral with respect to democracy and justice, the intrinsic value of the natural world, or the substance of human well-being (Solow, 1993). On a

9. Or, for another example, school sponsorship of sports linked to significant risk of traumatic brain injury, such as tackle football (Curren and Blokhuis, 2018).

typical weak theory, then (other variations are possible), sustainable development comes to nondeclining GDP over time.

Economic indicators of human well-being have obvious relevance as metrics of sustainable development. Advocates of strong sustainability tend to be skeptical, however, that all natural resources can be substituted in this way, or substituted effectively or without damage, ultimately, to human well-being or other important values. Therefore they propose that genuinely sustainable development would conserve, or avoid depleting, high-value stuff or natural resources, not least that upon which it relies (Norton, 2005, chap. 8). What is to be sustained is availability of high-value stuff for continued human access or use, and success or failure is to be measured using methods like ecological economics, systems ecology, ecological footprints, planetary boundaries, or environmental ethics or policy science. Strong sustainability thus involves passing on what Bryan Norton has a called a "structured bequest," one containing a particular package of goods—those enjoyment of which is regarded as fundamental to human well-being—not only economic wealth or technology (Norton, 2005, chap. 8).

The contrast between weak and strong sustainability is not dichotomous. It is possible to advocate more or less strong sustainable development, depending upon how the values of a growth economy, on the one hand, and natural resources and social goods like democracy and justice, on the other, are reconciled. ESD's ecumenical, three-legged stool ideal has the merit of singling out these core values and highlighting that they play different roles in sustainable development.

It has the drawback, however, of potentially appearing to suggest that those values easily coexist. The values of people, profit, and planet are all important, but they must be reconciled, that is, brought into harmony, and not simply maximally pursued separately, since some ways of pursuing them threaten the others (Schlottmann 2008).

The industrialized way of life that emerged from the Enlightenment West has yielded incredible rises in average wealth, health, and life expectancy across the globe. It also yielded colonialism, global North/South inequality, the climate change problem, and species mass extinction. Reconciling the core values of sustainable development will involve taking a controversial position on how we should respond to such difficult and complicated issues, as well as on deep philosophical issues such as the substance of human well-being and its measurement, facts that tend to be suppressed by the slogan "people, profit, planet."

For the Aristotelian, what is ultimately of value is flourishing and virtue. Economic goods are background external goods and have value only to the extent that they facilitate the exercise of virtue and achievement of the best things in life. Thus any sustainable bequest to future generations must be so structured as to support the development and exercise of virtue as part of its very point. Therefore, in any case where they are in conflict, the values of people, planet, and profit are to be reconciled so that the profit goal is always subordinate to the other two. Moreover, no purely economic metric of sustainability could be acceptable. A metric of well-being capturing the best intrinsic goods of human life—or else some process for making determinations concerning such goods (this is Norton's preferred solution)—must be used (Norton, 2005).

Aristotelian frameworks for conceptualizing and evaluating well-being—and therefore sustainable development—might be developed in a variety of ways. Some will have a closer kinship to strong sustainability, containing for instance a robust commitment to human–nonhuman animal relationships, or to aesthetic appreciation of nature, as part of the human good (Nussbaum, 2000). On such views, nonhuman nature is not substitutable with respect to human good because that good consists in part in relationship to nonhuman nature. This chapter is not the place to consider the many variations an Aristotelian view might take, nor address every difficulty such approaches might face. I address next the worry that there are no universal standards for virtue that could serve as guides for dealing with such issues.

AVE Helps Organize and Elaborate the Aims of ESD

Given the transformative intent of ESD, it is remarkable that its aims have not more often been articulated in terms of the virtues, including moral and intellectual virtues. ESD aims to shape learners' beliefs, values, attitudes, and behavior—in other words, what sort of person they will be—not just teach them *about* sustainability or impart merely academic skills which they may or may not go on to deploy in later life. Virtues define the sort of person one is and might overlap certain (sets) of ESD's transformative aims. When they do, these aims can be highlighted, associated, and elaborated by referring to the relevant virtues in their description, as well as linked to further priorities for ESD that emerge from a virtue ethical approach to sustainability.

Take for example the respect or democracy/peace aims. Relevant to these aims are the virtues of cosmopolitan justice and attachment, such as love, justice, fairness, goodwill, cooperation, or toleration. Appending these virtues to the statement of the democracy and peace aim, for instance, results in something like the following:

> Teach students to be just, fair, cooperative, and tolerant through a method that is participatory pedagogically and in purpose, so that learners participate in curricular decisions and can be committed to promoting democracy in a society without exclusion and where peace prevails.

This restatement highlights the transformative intent of ESD. These same virtues have obvious relevance to ESD's respect aims, too. Thus pairing the virtues to the aims highlights connections between them.

In one of the only treatments to note the relevance of the virtues, Randall Curren and Ellen Metzger, for example, have argued that teaching the Four Cardinal Virtues (justice, temperance, courage, and wisdom) can contribute to achieving many of ESD's goals. By way of teaching wisdom—they call it "good judgment"—for example, teachers can contribute to helping students "to understand matters of sustainability and human well-being, distinguish important truths from inconsequential distractions and propaganda, critically assess prevailing norms and practices, assess risk accurately, and think creatively about how to live flourishing lives in ways consistent with sustainability" (2017, p. 68). Thus teaching good judgment would help promote ESD's complexity and moral reasoning aims.

A virtue of good judgment is somewhat obscure, and it is not entirely obvious how acquiring this virtue should contribute to the learners achieving so much. We can unpack Curren and Metzger's thought, however, by decomposing good judgment into related sustainability virtues and seeing their relevance to various of ESD's ends. Take humility, a virtue of proper testimonial credulity, for example. Humility has been understood as the profile of dispositions manifesting due appreciation of one's limitations, and limited importance, as an individual and human being (Aikin and Clanton, 2010; Ferkany, 2015; Snow, 1995). This has been said to involve many things: being relatively unimpressed with one's own admirable qualities or talents, including as a knower or epistemic agent (Garcia, 2006); maintaining nonhierarchical and reciprocal relationships with others in which one avoids being very dismissive of the interests and opinions of others (Tiberius and Walker, 1998); declining to arrogate to oneself things to which one is not entitled (Roberts and Wood, 2007); even seeing humans as of limited significance from a cosmic point of view (Snow, 1995).

These dispositions are important aspects of a virtue of good judgment. They also have clear links to several ESD aims, including the respect and critical thinking aims. People who are disposed to maintain nonhierarchical and reciprocal relationships and see humans as of limited significance from a cosmic point of view are more likely, for example, to respect the human dignity of others and to "respect and care for the greater community of life in all its diversity." Being relatively unimpressed with their talents and powers as epistemic agents, they are also more likely to deliberate with others in a more open-minded and cooperative way and to be willing to subject their values to critical scrutiny. In its intellectual (epistemic) form, humility is an important virtue of proper testimonial credulity. It involves a due appreciation of one's shortcomings as a knower, and corresponding habits of deferring to more knowledgeable people for credible information as appropriate. The failure of the American public to respond to the climate change problem has many causes, but failures of proper testimonial credulity in a large cross-section of the population are partly to blame.

This sort of exercise could be undertaken for many other sustainability virtues. In the consumption-rich context of developed nations at least, the importance of virtues of right attachment to people and places—temperance, frugality, modesty, or what has been called simplicity—is almost too obvious to state. For example, problems of happiness or life satisfaction *and* pollution and biodiversity loss alike are driven by habits and patterns of consumption.[10] Individuals having the virtue of simplicity have "a conscientious and

10. I assume without argument here that this is true, even though problems of pollution and biodiversity can be moderated by improvements in technology that enable increased production of food, fiber, and consumer goods at a reduced cost to land, air, and water. Ultimately these are problems rooted in habits and patterns of consumption that, in some cases, may swamp the effects of improvements in technology. This seems to be the case presently relative to the climate change problem, and it will likely be the case for other problems, such as tuna conservation, in the absence of strict policy controls on consumption.

restrained attitude toward material goods" involving a "greater and more nuanced appreciation for other things besides material goods, and also for . . . material goods themselves" (Gambrel and Cafaro, 2010, p. 90). They spend more time and energy on social or self-actualizing activities, and thus both avoid the hedonic treadmill of materialist consumerism and limit their contribution to unsustainable consumption.

Thus specific ESD knowledge, values, and behaviors can be captured in terms of virtue. Denominating them in virtue terms should also enrich our understanding of what is educationally required to achieve them through schooling. I turn to this in the next section.

AVE Embodies a More Robust Framework for Attitudes and Behavior Education

A third merit of AVE for ESD is that it entails a richer and more demanding curriculum and method of transformative education. Many are familiar with the following sort of experience. You are attending a demonstration event at a nature center, or an aquarium or zoo. Center educators trot out various animals, or walk you through places of particular natural beauty, while describing their many amazing qualities, lifecycle facts, or ecological features or niches. Throughout, and certainly by the end, is mention of various "disaster facts," for example, rates of extinction or habitat loss. You may be told that "we," or you specifically, must take action or change your habits of consumption, transportation, eating, or whatever if the species is to survive or be restored.

Substantial evidence exists that this sort of "disaster education" is ineffectual (Heberlein, 2012; Kollmuss and Agyeman, 2002). It fails to significantly influence people's beliefs, values, and attitudes or, where it succeeds in that, does not result in significant or lasting change to their behaviors.

Facts and information and experiences of this sort should have a place in ESD. But if it is to succeed where traditional environmental education has failed, ESD must be different. It must not consist of a series of one-off experiences nor assume that inspiring people to care and informing them of relevant facts is sufficient for enabling them to adopt new behavior. It must assume instead that deep attitude change requires sustained experience of the relevant goods, and that behavior change requires coupling knowledge to habits and skills that have to be acquired through routine practice in a supporting environment. It must in other words bring the principles of AVE elaborated in this chapter to bear in shaping ESD policy, curricula, and pedagogy.

AVE Highlights the Moral Dimension of ESD

A final merit of integrating ESD and AVE is that virtue language helps highlight the moral dimension of sustainable development and the relevance of the humanities and social sciences to sustainability. In North America, sustainable development concerns have largely not been widely taken up in standards movements outside of the sciences. Social studies (history, civics, geography, economics), for example, is a natural home for dealing with sustainable development issues. However, no mention of "sustainability" or "sustainable development" appears in the Common Core State Standards recently adopted widely in the United States, nor in standards promulgated for history, civics, geography, and even economics (!) by the National Council for the Social Studies (NCSS, n.d.).

Sustainability issues *are* addressed in the sister standards for the STEM areas, the Next Generation Science Standards. This is unsurprising given the sometimes technical nature of sustainability problems, such as climate change. However, the NGSS incorporate sustainability issues in a technocratic way, all under the rubric of engineering and giving the impression that they are "to be treated as engineering problems" and "delimited as a set of neutral, quantitative problems, biophysical rather than sociopolitical" (Feinstein and Kirchgasler, 2015, p. 134).

It is widely agreed that sustainability problems are not mere technical problems. Key documents outlining the mission of ESD in fact sometimes use virtue and vice language to make clear that sustainability is an interdisciplinary problem ultimately rooted in problems of human values and behavior. According to UNESCO, for example:

> The human element is now widely recognised as the key variable in sustainable development, both in terms of reasons for unsustainable development and in terms of the hopes for sustainable development. Human relationships based on naked self-interest (greed, envy or lust for power, for example) maintain inequitable distribution of wealth, generate conflict and lead to scant regard for the future availability of natural resources. On the other hand, relationships characterised by justice, peace, and negotiated, mutual interests lead to greater equity, respect and understanding. It is these qualities that will underpin strategies of sustainable development. (UNESCO, 2006, p. 15)

It is impossible to know precisely why the import of this message has not been more widely recognized in educational standards movements and documents. One hypothesis is that the language of attitudes and behavior change characteristic of ESD mission documents "scientizes" what is in fact a deeply moral matter. Behind this might be the fear that being explicit about the moral root of unsustainability would provoke opposition to ESD on legitimacy grounds, that is, the view that it is not morally legitimate for governments—even worse, governments besides one's own—to use educational institutions to shape people's morals given that doing so would condition citizens' consent to live under their policies.

Having dealt with the legitimacy problem elsewhere, I will largely set it aside here (Ferkany, forthcoming; Ferkany and Whyte, 2013). In a nutshell, dealing with it is a matter of morally educating in the right way, not omitting to morally educate. Institutions inevitably shape the learner's moral character, whether by design or not. Aiming to do so explicitly and by design can enhance the legitimacy of any specific program of moral education by subjecting its content to democratic oversight and control. Failing to do so actually worsens the legitimacy problem by undemocratically institutionalizing the moral status quo. In this way being direct about the place of the virtues in ESD can both highlight the moral dimension of ESD, the place of non-STEM learning in ESD, and ease concerns about the legitimacy of ESD.

DOUBTS AND CHALLENGES

Not everyone will be persuaded of the value of approaching ESD through the lens of Aristotelian virtue ethics or virtue education. One doubt is that virtues concern individual behavior, whereas sustainability problems are ultimately problems of coordinated collective action at regional and global scales. Curbing climate change and preserving certain species will require nations to make policy in coordination with other nations. In the absence of coordinated policy change, sustainability problems are also systemic, confronting individuals with prisoner's dilemma–like situations they could not be expected to solve alone, for example, whether to risk bike commuting in a community without safe bike lanes. Making a goal of the virtues in ESD would be the wrong strategy to solve this sort of problem.

So put, this worry is more a criticism of the transformative aim of ESD generally—which is couched in the language of individual attitude and behavior change—than of the present approach to it.

But set aside, if the intended conclusion is that virtues education is unnecessary or unhelpful to achieving the goals of ESD, the argument for it turns on a false dilemma: either sustainability is achievable through individual behavior change, and educating for virtue is all that is required, or it is not and virtue is unnecessary.

It is controversial just how important individual behavior is to sustainability. By some ways of thinking about, for example, the climate change problem, individuals can make a very large difference indeed, enough to warrant holding themselves responsible for climate change outcomes (Hiller, 2011). But by Aristotelian lights, coordinated collective action and individual behavior change require one another and are mutually supporting. Individual behavior would not be sufficient to solve any social problem, but neither can social problems be solved if citizens lack the virtues. Virtue is learned through habitual practice in a supporting environment, one providing the right experiences and where norms of virtue are respected by community members. Thus directing ESD educators to make a goal of virtue in their teaching would not be to direct them to omit teaching about the systemic causes of unsustainability, and certainly not to ignore institutional aspects of the learning environment. On the contrary, educators who property adhere to the demands of virtues education will attend more thoughtfully to the learning environment than teachers adhering to traditional approaches to attitudes/behavior education, such as disaster education.

A second worry some might have concerns Aristotle's essentialist conception of human beings and the sentiment that the virtues are universal. Some might believe that the virtues are relative and vary by culture, so that it would not be possible or morally legitimate to specify the virtues to be taught as a common aim of ESD. The virtues needed for sustainable development may vary by nation or region, too, so that it would be unhelpful, or maybe even counterproductive, to attempt to specify them; educators in nations/regions needing citizens with one set of virtues might then erroneously teach for another set.

This worry misconstrues the virtue ethics on offer here, and it also commits a "partners in crime" fallacy. *Aristotelian* virtue ethics does not in fact presuppose a contested essentialist conception of a human being, nor attempt to extract some putatively universal set of virtues from that conception. Rather Aristotelians construct an account of the virtues observationally, by way of studying the conditions of human life generally and the dilemmas about which everyone must make a decision, such as how to secure food and shelter or navigate the complexities of human sexuality, reproduction, and the family system. Experience of these dilemmas can then provide the basis for cross-cultural discussion, yielding a broadly shared set of values and virtues that is sensitive to the different demands people face given regional cultural and environmental conditions (Nussbaum, 1988).

The problem of relativism is also not unique to AVE. Any approach to the aims and methods of ESD must acknowledge the scope for variation in values cross-culturally. In this respect AVE is just another partner in crime with any other approach to the values of ESD. Existing UNESCO guidelines handle this problem typically by advising that member nations incorporate "the values inherent in ESD" in ways that are suitable given local traditions and the like. A similar approach might be taken to applying the framework of AVE cross-culturally to ESD.

However, if the worry is that AVE would be too imperialistic because it is too universalizing, this approach might fairly be accused of providing too little guidance. ESD is intended to contribute to helping nations coordinate action toward the end of global sustainability. As discussed earlier, coherence problems emerge for ESD's conception of sustainable development even bracketing issues of cross-cultural variation in values, and one merit of AVE is its contribution to resolving this problem. Arguably another is that virtues are relevant even where controversy plagues the right resolution of certain sustainability value disagreements (say those relating to the relative priority of development and environmental conservation); a resolution that everyone can accept, for example, will partly depend upon the parties exercising certain virtues, such as proper testimonial credulity and a sufficiently cosmopolitan sense of justice. Where the patriotism of leaders and citizens of democratic nations underwrites xenophobia or an unwarranted sense of specialness, or their arrogance prevents them from appreciating the testimony of science experts, right resolution is unlikely to be forthcoming unless by force or manipulation. In this way, virtues of limited patriotism and intellectual humility, for example, are virtues citizens of a globally sustainable world need. If so, these virtues can serve as proper ends of ESD cross-culturally without passing into any troubling form of cultural imperialism.

CONCLUSION

Transforming the values, attitudes, and behaviors of citizens has been a central goal of the sustainability movement since its beginning. It is remarkable that linkages between this aim, virtue, and character education have not more often been drawn. While major international efforts to promote ESD are only just coming to a close, a consensus is emerging that they have largely failed to yield widespread uptake.

Many factors are likely responsible for this, but partly to blame might be inconsistencies and vagueness in the conceptualization of ESD and its execution. Linking ESD to AVE can contribute to remedying this situation. Reading the aims of ESD through AVE helps bring coherence to the values of ESD; it highlights the moral dimension of ESD and the linkages between ESD aims and traditional school subjects; and it superimposes onto ESD a framework for attitudes and behavior education that is far more robust than traditional disaster education.

Despite these merits, attempting to integrate AVE into ESD will not be easy nor remedy every problem of ESD. Even where educators understand and value the twin goals of AVE and ESD, school-based efforts will face many practical obstacles. AVE is thought to work at its best when schools, and individuals within them, model the virtues. But it is partly the nature of sustainability problems that structures and models for dealing with them are lacking. Educators in state-run schools will likely also have to negotiate policies of institutional bureaucracies at cross purposes to AVE or ESD (such as curricular standards devoid of either virtue or sustainability concepts). Overcoming these obstacles may require teachers and administrators to creatively reimagine their business in unforeseeable ways (see, for example, those discussed by Wright, Chapter 10, this volume).

At the same time, in the United States currently character education itself is frequently misunderstood, its broader implications for matters such as school funding, extracurriculars, or standards and accountability especially not well appreciated.[11] Resistance to the ideology of sustainable development will also have to be overcome for such efforts to even get off the ground. Working to change broader social norms must go hand in hand with school-based efforts to transform the sustainability values and behaviors of citizens. This is both an opportunity and a challenge for sustainability educators, for there is only so much institutionalized education can do to produce lasting change when the norms of the broader culture are at odds with the lessons educators wish to impart. If the Aristotelian's understanding of the conditions for educating for virtue are correct, educators have a long way to go before they can expect their efforts

11. Discussion of the various misunderstandings of character education can be found in Ferkany and Creed (2014) and Kristjánsson (2013).

to yield citizens who can reliably be expected to exercise the sustainability virtues.

REFERENCES

Aikin, S.F., and C. Clanton. 2010. "Developing group-deliberative virtue." *Journal of Applied Philosophy*, 27(4), 409–424.

Annas, J. 2011. *Intelligent Virtue*. Oxford: Oxford University Press.

Carson, R. 1965. *The Sense of Wonder*. New York: Harper & Row.

Curren, R. 2000. *Aristotle on the Necessity of Public Education*. Lanham, MD: Rowman and Littlefield.

Curren, R., and J.C. Blokhuis. 2018. "Friday night lights out: The end of football in schools." *Harvard Educational Review*, 88(2), 141–162.

Curren, R., and E. Metzger. 2017. *Living Well Now and in the Future: Why Sustainability Matter*. Cambridge, MA: MIT Press.

Curzer, H.J. 2012. *Aristotle and the Virtues*. Oxford University Press.

Feinstein, N., and K. Kirchgasler. 2015. "Sustainability in science education? How the next generation science standards approach sustainability, and why it matters." *Science Education*, 99(1), 121–144.

Ferkany, M. 2015. "Is it arrogant to deny climate change or is it arrogant to say it is arrogant? Understanding arrogance and cultivating humility in climate change discourse and education." *Environmental Values*, 24(6), 705–724.

Ferkany, M. Forthcoming. "Environmental education." In *Routledge Companion to Environmental Ethics*, edited by B. Hale and A. Light. New York: Routledge.

Ferkany, M., and B. Creed. 2014. "Intellectualist Aristotelian character education: An outline and assessment." *Educational Theory*, 64(6), 567–587.

Ferkany, M., and K.P. Whyte. 2013. "The compatibility of liberalism and mandatory environmental education." *Theory and Research in Education*, 11(1), 5–21.

Gambrel, J.C., and P. Cafaro. 2010. "The virtue of simplicity." *Journal of Agricultural and Environmental Ethics*, 23(1,2), 85–108.

Garcia, J. 2006. "Being unimpressed with ourselves: Reconceiving humility." *Philosophia: Philosophical Quarterly of Israel*, 34(4), 417–435.

Gutmann, A. 1999. *Democratic Education*. Princeton, NJ: Princeton University Press.

Heberlein, T. 2012. *Navigating Environmental Attitudes*. Oxford: Oxford University Press.

Hiller, A. 2011. "Climate change and individual responsibility." *The Monist*, 94(3), 349–368.

Hills, A. 2015. "The intellectuals and the virtues." *Ethics*, 126(1), 7–36.

Hursthouse, R. 1999. *On Virtue Ethics*. Oxford: Oxford University Press.

IVA Administration. n.d. "Intellectual Virtues Academy." Accessed October 26, 2016. http://www.ivalongbeach.org/.

Kasser, T. 2002. *The High Price of Materialism*. Cambridge, MA: MIT Press.

Kohn, A. 1997. "How not to teach values: A critical look at character education." *Phi Delta Kappan, 78*(6), 428–439.

Kollmuss, A., and J. Agyeman. 2002. "Mind the gap: Why do people act environmentally and what are the barriers to pro-environmental behavior?" *Environmental Education Research, 8*(3), 239–260.

Kraut, R. 2007. *What Is Good and Why: The Ethics of Well-Being*. Cambridge, MA: Harvard University Press.

Kristjánsson, K. 2015. *Aristotelian Character Education*. London: Routledge.

Leopold, A. 1949. *A Sand County Almanac*. Oxford: Oxford University Press.

Martusewicz, R.A., J. Edmundson, and J. Lupinacci. 2011. *EcoJustice Education*. New York: Routledge.

NCSS (National Council for the Social Studies). n.d. *College, career, & civic life C3 framework for social studies state standards: Guidance for enhancing the rigor of K-12 civics, economics, geography, and history*. Silver Spring, MD: National Council for the Social Studies.

Nolet, V. 2009. "Preparing sustainability-literate teachers." *Teachers College Record, 111*(2), 409–442.

Norton, B. 2005. *Sustainability: A Philosophy of Adaptive Ecosystem Management*. Chicago: The University of Chicago Press.

Nussbaum, M.C. 1988. "Non-relative virtues: A neo-Aristotelian approach." *Midwest Studies in Philosophy, 13*, 32–53.

Nussbaum, M.C. 2000. *Women and Human Development: The Capabilities Approach*. New York: Cambridge University Press.

Orr, D. 1991. *Ecological Literacy: Education and the Transition to a Postmodern World*. Albany, NY: SUNY Press.

Pinchot, G. 1910. *The Fight for Conservation*. Garden City, NY: Harcourt, Brace.

Roberts, R.C., and W.J. Wood. 2007. *Intellectual Virtues: An Essay in Regulative Epistemology*. Oxford: Clarendon Press.

Schlottmann, C. 2008. "Educational ethics and the DESD: Considering trade-offs." *Theory and Research in Education, 6*(2), 207–219.

Snow, N. 1995. "Humility." *Journal of Value Inquiry, 29*(2), 203–216.

Solow, R. 1993. "Sustainability: An economist's perspective." In *Readings in Environmental Economics*, edited by R. Dorfman and N.S. Dorfman, 179–187. New York: Norton.

Stone, M.K., and Z. Barlow, eds. 2005. *Ecological Literacy: Educating Our Children for a Sustainable World* (1st ed.). San Francisco: Sierra Club Books.

Tiberius, V., and J.D. Walker. 1998. "Arrogance." *American Philosophical Quarterly, 35*(4), 379–390.

UNCED. n.d. "Agenda 21: Earth Summit—The United Nations Programme of Action from Rio." UNCED. http://www.un.org/esa/dsd/agenda21/index.shtml.

UNESCO. 2002. "Education for Sustainability From Rio to Johannesburg: Lessons Learnt from a Decade of Commitment." UNESCO. https://unesdoc.unesco.org/ark:/48223/pf0000127100.

UNESCO. 2005. "UN Decade for Sustainable Development: The DESD at a Glance." UNESCO. http://unesdoc.unesco.org/images/0014/001416/141629e.pdf.

UNESCO. 2006. "Framework for the UNDESD International Implementation Scheme." UNESCO Education Sector. http://unesdoc.unesco.org/images/0014/001486/148650e.pdf.

Watson, G. 1990. "On the primacy of character." In *Identity, Character, and Morality*, edited by O. Flanagan and A.O. Rorty, 449–470. Cambridge, MA: MIT Press.

Chapter 4

The Emotional Heart of Environmental Virtues

CHERYL HALL

What kinds of character traits must people have to live in ways that are environmentally sustainable and just, both individually and collectively? What virtues do they need to be "green"? Environmental virtue ethicists have argued for a variety of different virtues as necessary or useful for sustainability. In doing so, they have provided important justifications for such virtues and carefully explored their cognitive and behavioral characteristics. In many cases, however, they have paid relatively little attention to the emotional characteristics of the virtue(s) under discussion. In this chapter, I argue that environmental virtue ethics ought to pay more attention to emotions, because virtues cannot exist without their contribution. As with all virtues, environmental virtues depend on particular ways of emotionally engaging with oneself and the world, ways of engaging that must in turn be cultivated using advanced skills in working with emotions. So if we wish to understand how to develop and sustain the virtues we need for a more sustainable and just world, then we need to understand the emotional components of environmental virtues and how to cultivate them.

Cheryl Hall, *The Emotional Heart of Environmental Virtues* In: *The Virtues of Sustainability*.
Edited by: Jason Kawall, Oxford University Press (2021). © Oxford University Press.
DOI: 10.1093/oso/9780190919818.003.0004

The chapter begins with an argument of the need for environmental virtues alongside the need for structural changes. Next I provide an overview of some of the many environmental virtues that have been proposed to date. Turning to my central argument, I note that attention to emotions remains fairly minimal in this literature. To begin remedying this oversight, I discuss three ways in which emotions lie at the heart of environmental virtues. The chapter concludes with some brief thoughts on the implications of the argument.

In the context of a volume focused on virtues of sustainability, I should note from the start that I use the term "sustainable" to mean that which supports, protects, preserves, or restores the ongoing well-being of the web of life on Earth and its conditions for flourishing. In speaking of "the web of life" rather than "nature" or "the environment," I aim to avoid erroneously conceptualizing humans as somehow clearly separate and distinct from the rest of nature or our surroundings, as the dominant connotations of "nature" and "the environment" tend to do. Because humans are in fact deeply interconnected with the rest of nature in uncountable ways, to protect the web of life is to protect the well-being of human life together with other forms of life on Earth and the various material cycles that support all life. I take the key ecological point to be that nothing exists in isolation, that countless strands ultimately connect everything in existence in a complex, interdependent web (Thiele, 2011). Given such interdependence, the viability of each part of the web depends in great part on the viability of the rest. The ultimate aim of "sustainability," then, is to protect the integrity, strength, and resilience of the web of life itself.

One further note on terminology bears mentioning. Since the Brundtland Report issued by the U.N. World Commission on Environment and Development in 1987, it has become exceedingly common for the word "sustainability" to be intentionally or unintentionally conflated with a specific concept of "sustainable development." Within the framework of this concept, the three values of social equity, economic growth, and environmental protection (or "people, profit, and planet")[1] are considered equally important and indeed mutually constitutive when properly aligned (World Commission on Environment and Development, 1987). I do not have space to make the argument here, so I can only assert that in my understanding the capitalist economic system that currently dominates the globe inescapably requires not just profit and growth but continued inequitable accumulation of capital and unlimited exponential growth in production and consumption, which inevitably undermines both social equity and environmental protection, making "sustainable development" as currently conceived a contradiction in terms. Accordingly, in my usage "sustainability" is decidedly not synonymous with "sustainable development." That being said, I do not believe the web of life on Earth can be effectively preserved without just and equitable relations within and across current and future human and nonhuman populations. In the argument that follows, then, I presume that an environmentally sustainable future is tied to values of justice and equality.

THE NECESSITY OF ENVIRONMENTAL VIRTUES

It is important to be clear that individual virtuous actions are not sufficient on their own to create a more sustainable future, since major

1. A slogan highlighted by Ferkany (Chapter 3, this volume). Ferkany argues that one advantage of an Aristotelian virtue model for sustainable development education is that it can help prioritize the values of people and planet over that of profit "in any case where they are in conflict" (C3.P36). I agree with the priority, but contend that the value of profit will in fact always conflict with the values of people and planet.

contemporary environmental problems such as climate change, biodiversity loss, soil depletion, and widely dispersed toxic chemicals spring in great part from the political, economic, and material structures that are currently dominant in the human world.[2] These structures include the aforementioned global capitalist economy; a global political assemblage of largely competitive territorially based nation states beholden to economic growth and (if democratic) paced by short electoral cycles that privilege short-term goals over long-term sustainability; and the ongoing development of environmentally damaging forms of infrastructure and technology that make it increasingly difficult to pursue greener paths (for example, urban design and transportation systems that privilege and thus perpetuate personal automobile use). Effectively resolving the biggest environmental problems currently facing the world will not be possible without modifying these political, economic, and material structures.

Yet virtues of character are still necessary for this task, because structural solutions are themselves not possible without a reasonable percentage of people in any given human community having reasonably stable internal dispositions to act in environmentally sustainable ways. It's true that people can be externally pressured or influenced to act in green ways through a number of methods, ranging from laws, fines, and other kinds of force to incentives, design, and other kinds of nudges to information, norms, and other kinds of education and enculturation.[3] But none of these methods can do the job on their own, either. Absent significant popular support, legal prohibitions and requirements are politically costly and quite likely to generate backlash (Ockwell, Whitmarsh, and O'Neill, 2009). Indeed, the lower the support among a populace, the more force is required

2. For further support of this argument see Bendik-Keymer (2012), Clowney (2014), Kawall (2018), Ockwell, Whitmarsh, and O'Neill (2009), Treanor (2010), and Vogel (2012), among others.

3. For more discussion of the potential role of what I call "context design," see Hall (2016).

to make people follow the law, and the greater the force employed, the greater the potential resistance. Meanwhile, nudges may be less politically costly, but only if aimed at changes that are relatively small or easy for people to make, such as lowering thermostats a bit or recycling more. Unfortunately, at least in the case of advanced industrialized societies, human beings are currently consuming natural resources and emitting greenhouse gases and other pollutants at rates that must be reduced by 40%–100% in order to become sustainable.[4] And while the changes required may well make for a better way of life overall, at least some of them will not be particularly easy to nudge people into making.[5] Finally, the method of changing norms can certainly make a big difference, but doing so from the top down is difficult, and it is well established at this point that simply providing people with more information or guilt-tripping them does little to change their views or behavior. All of these external ways of affecting how people act will be part of the solution, then, but none of them—not even the combination of all of them—can do the whole job of moving advanced industrialized societies in particular toward a more

4. The average US resident uses approximately five times as many resources and waste sinks as the planet can support for every human being ("Global Footprint Network", n.d.). To achieve a world that is both sustainable and just, US residents need to reduce their ecological footprints by approximately 80%. While the ecological footprints of most European residents are smaller, reductions still fall in the 40%–50% range. Meanwhile, necessary reductions in carbon emissions are even steeper. A 2018 IPCC Special Report identifies necessary global reductions of about 50% by 2030 and 100% by 2050 in order to avoid warming of 1.5°C above preindustrial levels (Levin, 2018). Leaving carbon emissions space for lesser-developed countries to catch up on industrialization would require advanced industrialized countries to reduce their emissions even faster.

5. For example, in spite of advanced technological capabilities, current prospects for powering airplanes that can transport large numbers of people across long distances at high speeds without burning fossil fuels are slim (Jacobs, 2016; Plumer, 2016). So it may just be that the extraordinarily cheap and easy long-distance travel currently provided by airplanes to billions of passengers each year—over 4 billion globally in 2017 (Wood, 2018), with demand doubling every 15 years (Qualman, 2017)—is not conducive to a stable climate. While shifting the incentives involved in the cost and convenience of various forms of travel will surely influence people to fly less, neither will it automatically transform the attitudes, values, commitments, and habits of the vast numbers of people now accustomed to plane travel.

sustainable future. Moreover, at least in democratic regimes, most of these external factors still require some significant number of people willing to advocate for them and to support political representatives, educators, and other leaders who will put them into place.

To a considerable extent, then, a more sustainable future will require some significant number of people to have some *internal* motivation to pay attention to, learn about, and prioritize the value of living sustainably. This point holds not only for ordinary citizens but for political representatives and others with power to make decisions affecting the sustainability of a society (Kawall 2018), as well as for anyone who can influence the culture of a society: what it perceives, understands, counts as normal, valorizes as good, and so forth. But this motivation to pay attention to and prioritize the value of sustainability is at least somewhat difficult, especially in the context of an unsustainable culture, economy, and political world. It requires seeing what is obscured by the hegemonic worldview. It also requires prioritizing long-term over short-term interests, taking an expansive rather than narrow perspective, and valuing nonmaterial as well as material interests, among other challenges.

This is exactly the kind of context in which virtues of character matter. By "character," I mean a relatively consistent ethical disposition comprised of a person's usual patterns of perception, thought, feeling, and action. What does a person tend to notice in their inner and outer world? What do they tend to ignore or dismiss? What habits of thought and feeling do they have? How do they ordinarily act (or not) in a given situation? And how do all of these things shape their ethical relationship to the world? People are messy, complicated creatures who are often contradictory and changeable, especially when circumstances change. We are never perfectly unified or consistent. That being said, most of us still tend to develop patterned ways of being in the world over time. When such patterns have moral consequences, we generally call that character. By "virtues" of character, I mean dispositional traits to perceive, think, feel, and act in ways that are at least arguably admirable as excellent. For the purposes of this argument, I leave aside the question of whether there is any objective standard (including the naturalistic one of "human flourishing") by which "excellence" can be judged. My concern in this chapter is not to evaluate the status or justification or relative worth of any of the environmental virtues proposed to date; I take it that there will be broad enough agreement that most if not all are arguably excellent human traits, particularly under current circumstances of grave environmental threat. Proceeding from this assumption, my concern is rather to investigate something of the emotional composition, texture, and development of any of these virtues. Because sustainable ways of life depend in part on people's internal commitments to prioritize them, and because people's commitments are shaped at least in part by their character, it makes sense to explore not only which character traits will support environmental sustainability, but the full nature of those traits and how to cultivate them.

PROPOSED ENVIRONMENTAL VIRTUES

Many kinds of traits have been proposed as environmental virtues. Most—though not all—are general traits that have long since been claimed as virtues in some other context, now submitted as relevant to the specific good of human beings living in balance with the rest of life on Earth. Needless to say, in each case there is a corresponding environmental vice (or set of vices, if Aristotle is right that virtue is a mean between extremes) that the virtue is meant to counteract, although most authors focus significantly more on the virtue than the vice(s).[6] In what follows, I offer an extensive but nonexhaustive list meant to

6. Exceptions include Bendik-Keymer (2010), Cafaro (2005), Kawall (2018), and Norlock (2010).

provide an overview of the literature on environmental virtues. For the sake of surveying the terrain in an organized way, I have grouped these virtues into several categories. However, the lines I've drawn are both provisional and porous: these categories are not the only ones possible, and many virtues could be placed in more than one category anyway. Notably, several authors argue for a whole set of environmental virtues, suggesting that the virtues they advocate for are at least complementary if not connected. Setting aside these complexities, though, I see the following five kinds of environmental virtues being proposed.

The first and largest set are virtues in how people care *about* and *for* the rest of nature. Notice that "care" here has two components: it is both a perception and an action, both a way of *valuing* the rest of nature and a way of *behaving* in relationship with it. As I have argued elsewhere, the two parts of care are linked, for claims to value a thing that are not accompanied by any action on its behalf (when necessary and possible) are shallow if not insincere, while actions purportedly on behalf of a thing that are not accompanied by any belief in its value lack sufficient attunement and motivation to truly benefit it (Hall, 2019, p. 217). While some arguments for virtues in caring about and for the rest of nature place more stress on the perception of value and some place more stress on the action of caretaking, all at least implicitly presume the connection between the two. This category of environmental virtues has a number of subsets. The most obvious subset is what we might broadly call *love* for the rest of nature. Many other words have been used in addition to love, including *biophilia, care, reverence, thoughtfulness,*[7] *benevolence, compassion,* and *mercy* (Abbate, 2014; Bendik-Keymer, 2010; Clowney, 2013; Dixon, 2010; Ferkany, 2011; Frasz, 2005; Hall, 2019; Hulme, 2014; Hursthouse, 2011; Kawall, 2003; Lenzi, 2017). Each of these terms highlights

7. Bendik-Keymer's "thoughtfulness" does not mean "being full of thought" but rather "noticing and caring." In his argument, the term is quite close to "reverence" (2010, p. 77).

somewhat different qualities. For example, reverence for life emphasizes "treat[ing] living things as intrinsically valuable" (Kawall, 2003, p. 340) while environmental benevolence emphasizes "an active and consistent concern for the happiness, flourishing, health, interests or well-being of both human and non-human others" (Frasz, 2005, p. 125). Still, some form of love arguably underlies all of these virtues. Notably, love or care for the rest of nature need not present itself in directly "positive" form; rather it may take the outer form of *revulsion* (Abbate, 2014) or *anger* at its mistreatment.[8] Again, though, it is love, concern, and compassion for the mistreated that underlie virtuous revulsion or anger at their mistreatment. A rather more specific form of love is *patriotism*, which Cafaro argues is an environmental virtue insofar as it consists of "devotion to particular places" as well as to the people who live there (2010, p. 185). Another subset within the category of caring about and for the rest of nature includes the virtues of *respect, acknowledged dependence*, and *gratitude* (Hannis, 2015; Jain, Chapter 5, this volume; Loder, 2011). Here the focus is especially strong on a deep understanding of and appreciation for what the web of life provides us as human beings. On the other hand, the virtues of *responsibility, conscientiousness, integrity, stewardship, Ahimsa (nonviolence)*, and *justice* focus more on what human beings are obligated to provide to and for the web of life (Hulme, 2014; Jain, Chapter 5, this volume; Mason, 2014; Thompson, 2012; Vanderheiden, Chapter 9, this volume).[9] A final subset identifies the virtue to consist precisely

8. Abbate primarily argues for the virtue of compassion, but she repeatedly references revulsion, aversion, and disgust as the kinds of reactions that a virtuous person would feel in response to the unnecessary pain and suffering of other animals. I include this example especially to highlight that the emotions involved in environmental virtues are not limited to those conventionally labeled as "positive," a point that will be reinforced later when I argue that all emotions track values and thus moral relevance. I thank Jason Kawall for drawing my attention both to the question of "negative" emotions in virtue ethics and to Abbate's article.
9. In the case of justice, it is especially important to remember that the web of life includes other human beings.

in the mutuality of give and take within this relationship, emphasizing *friendship* with the rest of life on Earth (Bannon, 2017).

The second set of environmental virtues are virtues in how people manage their appetites, desires, and egos in relationship to the rest of nature. With regard to appetites and desires, a number of authors argue for the virtues of *moderation* (or *temperance*) and *simplicity* in the acquisition and consumption of material goods (Cafaro, 2005; Dixon, 2010; Gambrel and Cafaro, 2010). Whereas classic arguments for moderation and simplicity usually focus on the corrosive effects of greed and gluttony on the human soul or on relationships between human beings, here the argument is rather more direct, since the more material goods human beings produce and consume, the more natural resources they deplete and/or pollute and the more they threaten the integrity of the web of life. With regard to ego, authors also argue for *humility* in how people view their relative importance and power as compared to the rest of existence, their abilities to fully understand the rest of existence, their abilities to effectively manipulate and control it, and their right to do so solely for their own interests (Cafaro, 2005; Hulme, 2014; Pianalto, 2013).[10]

A third set of environmental virtues concerns virtues in how people orient themselves in time: how they relate to the past and the future, and how this relationship affects their choices in the present. Norlock (2010) argues for the virtue of "preservative" *forgiveness* in response to the human legacy of environmental harm: a commitment to look forward rather than backward in order to avoid the ethical perils of excessive recrimination, pessimism, and despair about human action that can undermine the motivation for environmental activism. Thus, she sees the virtue of *forgiveness* about the

10. As an illustration of the way in which virtues can cross the lines between these categories, consider that the virtue of *acknowledged dependence* could easily fit alongside that of *humility*, while the virtue of *gratitude* for what Earth has already provided could easily fit alongside the virtues of *modesty* and *simplicity*.

past as necessary to the virtue of *hope* for the future, which in turn is necessary to the virtue of *ecocitizenship* in the present. A number of other authors also argue for the virtue of *hope* or *faith* in the prospect that it is still possible for human beings to live in better balance with the rest of life on Earth; that doing so can in fact bring benefits to human beings and not just deprivations; and that it is still worth acting toward this goal, regardless of current obstacles and any guarantees for success (Hulme, 2014; Mason, 2014; Thompson, 2010; Treanor, 2010). Relatedly, Kawall (Chapter 8, this volume) argues for *patience* as an environmental virtue: a willingness to accept that actions intended to help solve large environmental problems may not bear fruit for a long time, perhaps for many generations, and should be undertaken even so. Once again, a particular kind of relationship toward the future—here, releasing the need for immediate results or even clear indications of progress toward one's goal—is identified as necessary to support persistence in taking action in the present.

A fourth set of environmental virtues draw especially upon (without being limited to) intellectual capacities: that is, abilities to pay attention, to come to know and understand, to imagine, to analyze, and to solve problems. These virtues include *creativity, open-mindedness, imagination,* and *practical wisdom* (Hulme, 2014; Mason, 2014; Wright, Chapter 10, this volume). Authors who argue for environmental virtues such as these emphasize the importance of human intellectual abilities in understanding the existing relationship between humans and the rest of nature, evaluating the implications of that relationship, coming up with solutions for more sustainable relationships, and developing paths toward achieving such solutions. In all of these cases, perception, intellectual flexibility, judgment, and careful reasoning are key foundations for knowing how best to act environmentally.

The final set of environmental virtues are virtues by which human beings are able to act in environmentally sustainable ways, individually and collectively. These virtues often build on many of the virtues identified earlier, yet they also draw upon some additional traits. A number of authors argue that living more sustainably requires what Treanor calls *political virtues*, "virtues possessed by an individual that express themselves in the political arena" (2010, p. 10) or even what Clowney calls *collective virtues*, virtues that are possessed by a group as a group (2014, p. 318). The key argument for either political or collective virtues is that significant environmental problems stem from the accumulated actions of a great many people, whose actions are in turn shaped within political, economic, technological, material, and cultural contexts. As such, these problems cannot be solved solely by uncoordinated individual actions, however virtuous. Rather, they require individuals to engage carefully and actively within and between their communities in order to advocate for organized collective approaches to the problem. Among the political or collective virtues recommended are *cooperation*, *participatory virtues*, and *ecocitizenship*, understood as "environmental activism and dedication to policy change" (Ferkany and Whyte, 2012; Hartman, Chapter 7, this volume; Norlock, 2010, p. 30). Two additional virtues important in being able to act in environmentally sustainable ways are *courage* to stand up for environmental rights or against environmental wrongs (Fredericks, 2014) and *environmental skill*, a concrete, embodied, practical, experiential "knowing-how" to engage with the rest of nature (Coeckelbergh, 2015).[11]

EMOTIONS AND ENVIRONMENTAL VIRTUES

Again, I presume that most if not all of these proposed environmental virtues can potentially play at least some role in helping people

11. *Environmental skill* is somewhat similar to *practical wisdom*, and both show how porous the line between the categories of "knowledge" and "action" can be.

to create more sustainable societies. There is plenty of room for further debate about which of these virtues might be most important to prioritize, how various virtues might be related, whether there are any tensions between them, and so forth. What I want to emphasize, though, is that emotions are central to the development and exercise of all of them. There is a common tendency in both contemporary academic discussions and ordinary conversation to see virtues as primarily an achievement of the human capacity for reason (and/or will), overlooking the achievements of the capacity for emotion that are also involved. Indeed, often enough virtues are seen as an achievement of our capacity for reason *over and against* our emotions, rather than as a collaborative enterprise of both. This tendency is especially ironic in philosophical literature given how foundational Aristotle's theory of virtue is to the tradition of virtue ethics and how central emotions are to his theory. Nevertheless, many arguments on behalf of virtues fail to explore the emotional components of virtue in any depth or detail. While there are some notable exceptions, this includes many arguments on behalf of environmental virtues.[12] In some cases authors overlook emotion entirely. More often they acknowledge in passing that an environmental virtue includes emotional elements or even is (in part) an emotion itself but then provide little discussion of the phenomenological experience of the emotions involved, their texture and composition and valence and intensity, their relationship to accompanying perceptions and sensations and thoughts, what it might take to cultivate and then sustain these emotions, and so forth. As I have already suggested, I think it's a substantial mistake to pay so little attention to emotions, because virtues cannot exist without their contribution. If this is true, then overlooking the emotional components of environmental virtues

12. Exceptions include Abbate (2014), Bendik-Keymer (2010), Coeckleberg (2015), Hursthouse (2007, 2011), Mason (2014), and especially Loder (2011).

will prevent us from understanding how to develop and sustain the virtues we need for a more sustainable world.

There are at least three ways in which emotions play an important role in environmental virtues. First and most obviously, some environmental virtues are centrally characterized by key emotions directed toward the web of life on Earth. Second, although less obvious, all environmental virtues (as with all virtues in general) are in fact partly constituted by a variety of emotional dispositions and practices. Third, a general facility or skill in working with emotions is necessary in order for any environmental virtue (as with any virtue in general) to develop and be sustained. I discuss each of these ways in which emotion plays a role in environmental virtues in turn.

ENVIRONMENTAL VIRTUES CHARACTERIZED BY EMOTIONS CONCERNING THE WEB OF LIFE

It is not hard to see that some environmental virtues are centrally characterized by key emotions concerning the web of life on Earth. The most recognizable examples are in the first set of virtues listed earlier: love, care, reverence, compassion, and gratitude for life on Earth.[13] Here the emotional component of the virtue is quite prominent, making it fairly apparent that the virtue just is (in part) a kind of appreciation and concern (caring about), combined with action based on that concern (caring for). The virtue is not simply in *thinking* or *perceiving* or *believing* that the web of life matters or has value or should be appreciated and cared for as a matter of rational principle;

13. While the emotional component is perhaps less noticeable in the cases of moderation, simplicity, and humility regarding the web of life, these environmental virtues arguably fit in this category as well.

it is also in actually *feeling* the importance or value of the web of life in one's heart, actually *experiencing* the emotions of love and gratitude for life, actually *caring* about life in the flesh. That is to say, the virtue is not an abstract or detached intellectual position but an embodied, heartfelt attachment leading to concrete action in the world. And it is precisely because the virtue requires tangible action in the world (where necessary and appropriate and possible) that it is so important to emphasize its emotional character, for abstract, detached intellectual positions do little to motivate people to act, while heartfelt attachments provide powerful inspiration to do just that.

Having emphasized that forms of appreciation and concern for the web of life are experiences that are emotional and not *just* cognitive, though, I hasten to add that they are certainly cognitive as well. Love, care, reverence, compassion, and gratitude are not random urges or inexplicable appetites or mindless feelings. If a person loves, reveres, or is grateful for something, it's for a *reason*: because they judge that thing to be *worthy* of love, reverence, or gratitude. They believe it to be good or valuable in some way, and their emotional appreciation and concern for it are tied to this assessment. The judgment of value may be more or less explicitly identified, more or less carefully considered, but without it, the person would not love or revere or be grateful for that thing. In the case of forms of love for the web of life on Earth, love registers the value of the web of life and a corresponding commitment to support it. Such virtues are both heartfelt *and* intelligent, both embodied *and* conceptual.

EMOTIONAL CONSTITUENTS OF ALL (ENVIRONMENTAL) VIRTUES

So virtues of appreciation and concern for the web of life consist of at least three connected components: emotion, cognition, and action.

That all three elements—emotion included—are part of the virtue may be relatively easy to see in the case of these virtues, but the larger lesson is that all virtues are actually composed of all three elements. This lesson about the nature of virtue goes back at least as far as Aristotle (Cooper, 1999; Kosman, 1980; Nussbaum, 1994; Sherman, 1995, 1999b, 2000).[14] For Aristotle, virtue is a disposition of both feeling and acting as cultivated and refined through practical wisdom. Crucially, in his view it is "not simply how one acts *in light of* one's feelings that one is said to be virtuous" but that feelings are "*part of the concept of that virtue* considered as a disposition" (Kosman, 1980, pp. 100, 109, my emphasis). In other words, the courageous person is not someone who has learned to act courageously *in spite of* feeling fear. Rather, the courageous person is someone who has learned at a deeper level how to actually feel fear when appropriate and not feel fear when not appropriate, thus cultivating an emotional disposition that will help them to act courageously rather than brashly or cowardly (1980, p. 108). So it is a matter of becoming the kind of person who will feel the right emotions in the right way under the right circumstances for the right reasons and then act accordingly. To turn the point around, even if a person acts in the "right" way, for Aristotle the act is not virtuous unless done for/from the right reason/feeling. A person who fought against injustice but did so out of resentment and vengefulness rather than a commitment to justice would not be a virtuous person for him. Whether one accepts Aristotle's argument in its entirety or not, it seems reasonable to say that virtue is most consistent and reliable when emotion, cognition, and action align. Vengeful, resentful people are unlikely to do the right thing for long, just as those with no emotional investment are unlikely to do

14. I emphasize Aristotle here because of his importance to the Western tradition of virtue ethics, not because he is the only one to make the argument.

anything in the first place even if they "know" intellectually that they should.

Contemporary theorists further support this understanding of virtue by identifying integral connections between emotion, reason, ethical judgment, and ethical action (Calhoun, 1984; de Sousa, 1980; Nussbaum, 2001; Sherman, 1999a, 1999b; Solomon, 2003). I noted earlier that love contains an implicit cognitive judgment of the worthiness of the beloved object. Again, that a judgment is involved may be relatively easy to see in this case, but the point is that all emotions actually entail "appraisals" about objects and situations (Sherman, 1995, p. 4). All emotions contain implicit if not explicit cognitive judgments about value and circumstance. For instance, anger registers the assessment that a circumstance is unjust because something one values has been undeservedly injured or withheld, fear registers the assessment that a circumstance is unsafe because something one values is threatened, and sorrow registers the assessment that a circumstance is deficient because something one values has been lost. Emotions—including those we consider "positive," those we consider "negative," and everything in between—are "modes of registering value (e.g. what is dangerous or beneficial, needy or attractive, disturbing or insulting)" (Sherman, 1995, pp. 2–3). Because they track what we value and how we judge the appropriateness of the circumstances, emotions are also "modes by which we track moral relevance" (Sherman, 1999a, p. 297). A combined nexus of thought and feeling is what helps us to identify situations of suffering, violence, and injustice and what motivates us to respond to those situations. Without appraising emotions such as compassion and indignation, "we are simply insensitive to much crucial moral data and are morally bumbling in our interactions with others and our understanding of ourselves" (1999a, p. 298). A virtuous way of being in the world thus depends on particular forms of connected emotions, cognitions, and actions.

Notably, if emotions are integral to virtue, then they are included within the realm of moral responsibility. One might object that human beings cannot be held responsible for our emotions because they are experiences that just happen *to* us. Such a protest is certainly long-standing and widespread, but it misses the myriad and profound ways in which emotions do not just "happen" to us. In fact, we work to construct and reconstruct our emotions on a nearly continual basis.[15] We encourage, discourage, channel, sublimate, project, soothe, allow, deny, reinforce, repress, intensify, label, intellectualize, and in many other ways cultivate and manage our emotions—all within the context of people and cultures around us also working to cultivate and manage our emotions. That often we are not fully aware that we are doing this does not mean that we are not doing it, or that we cannot become more aware of how we are doing it. The process of increasingly conscious awareness and cultivation of emotions begins in infancy and is as significant a part of human development as the increasingly conscious awareness and cultivation of other capacities such as those for logic, language, and physical coordination (Sherman, 1999a). In the best circumstances, the process leads to adults with high levels of emotional competence, but even under the worst circumstances we consider people to become increasingly more "responsible" for their emotions as they grow from babies into adults. None of this is to say that any of us achieves some kind of perfect "control" over emotions—but then neither do any of us achieve

15. Recent research in neuroscientific psychology suggests that emotions are not just cultivated or managed but entirely constructed in the brain using culturally specific learned categories to interpret sensations and perceptions (Barrett, 2017). That is to say, they are conceptual (and social) all the way down. The basic valence of a feeling being "pleasant" or "unpleasant" is universally shared across all cultures; beyond that, the particular substance of what physical sensations, facial expressions, sounds, thoughts, meanings, and values are mapped onto what emotion categories can vary widely across cultures and even individuals. What this means is that "emotions are not reactions *to* the world; they are your constructions *of* the world" (2017, p. 104).

perfect control over our thoughts, speech, or actions, either. With regard to all of these capacities, we hold people responsible within the context of their past and present conditions and abilities (or so we should) and admire the excellence of those who exhibit high levels of facility.

If all virtues are composed of a nexus of emotion, cognition, and action, then all environmental virtues are composed of a nexus of emotion, cognition, and action. To get some idea of how emotions in particular help to constitute environmental virtues, consider a sampling of one virtue or set of virtues from each of the categories discussed earlier. In each case, the virtue requires emotionally registering something (or several things) as valuable as well as engaging in ongoing emotional work to cultivate and sustain a disposition that can support what is valued. So the environmental virtue of taking *responsibility* for the web of life and its conditions for existence requires emotionally registering the value of the web of life: that is, not just believing but genuinely *feeling* that it is deserving of committed responsibility.[16] It also requires doing the ongoing emotional work involved in maintaining the focus and motivation necessary to keep this commitment in the face of competing obligations, external challenges, changing circumstances, confusion about how to proceed, despair about the prospects of success, and strong temptations to absolve oneself (or selves) of any accountability in the face of collectively caused problems. Or to put the point another way, it takes emotional energy and discipline to "be willing and able to do one's best . . . to ensure that things come out right" (Thompson,

16. Thompson (2012) would emphasize that this responsibility takes on a new form now that human beings have modified planetary systems (such as the climate) so extensively that we must increasingly actively manage them to prevent their collapse. Accordingly, in this case the process must shift from emotionally registering the value of some autonomous, pristine, wild "nature" to being able to emotionally register the value of a fully hybridized human/nature system.

2012, p. 216). The environmental virtue of *humility* with respect to the rest of nature requires comprehending and embracing the value of other living beings and systems on an emotional as well as intellectual level: truly appreciating myriad forms of strength and beauty and intelligence and complexity beyond the human. It also requires doing the ongoing emotional work of coming to terms with a certain amount of inescapable vulnerability and helplessness and ordinariness rather than continuing to fantasize about being uniquely powerful and knowledgeable and important and deserving in order to feel safe and special. The environmental virtue of *forgiveness* of environmental sins requires emotionally recognizing the value of the healing, reconciliation, and preservation that a focus on forgiveness instead of condemnation can provide. It also requires doing the ongoing emotional work involved in letting go of feelings of blame and bitterness and cultivating in their place feelings of compassion, civility, and faith in the possibilities that people can do better. The environmental virtues of *creativity* and *open-mindedness* in inventing new solutions for sustainability or new forms of environmental activism require emotionally acknowledging and prioritizing the values of imagination, playfulness, receptivity to inspiration, and nonconformity. They also require doing the ongoing emotional work involved in letting go of habitual attachments to familiar and comfortable ways of thinking and being in the world, tolerating the discomfort of staying open to what is new, uncertain, and/or previously considered disagreeable,[17] remaining relaxed and flexible in one's perspective, holding inner and outer critical voices in abeyance long enough to allow inspiration to flow freely, and committing to process over product even in the face of dire need for solutions. Finally, the *participatory virtues* necessary

17. Wright gives the example of having to overcome the emotional reaction of disgust in order to be able to embrace a sustainable solution of using sterilized human waste for fuel (Chapter 10, this volume).

to deliberate with others about "wicked" environmental problems all require emotionally embracing the value of active, consistent, cooperative membership in a community (among many other values). The list of proposed participatory virtues is extensive, including empathy, sincerity, self-confidence, humility, attentiveness, patience, resilience, and reasonableness (Ferkany and Whyte, 2012). It should be clear by now that cultivating such virtues requires ongoing emotional work involved in managing feelings of fear, doubt, frustration, anger, pride, despair, fatigue, and many more that can undermine the focus and energy and dedication and openness necessary to work through disagreements about how to handle difficult challenges collaboratively. In all of the environmental virtues just discussed, the need for people to *think* and *act* in certain ways in order to embody the virtue may be more immediately obvious, but the need for people to also *feel* in certain ways in order to embody the virtue is no less crucial.

EMOTIONAL FACILITY AS A BACKGROUND CONDITION FOR ALL (ENVIRONMENTAL) VIRTUES

Some environmental virtues are centrally characterized by emotions concerning the web of life on Earth, then, and all environmental virtues depend on various kinds of emotions in order to be the virtues that they are. The final point is that developing and sustaining any of these emotions requires a sufficient level of emotional facility or intelligence. Again, this point holds for environmental virtues because it holds for all virtues. Virtues depend on the cultivation of certain kinds of emotions and the transformation of other kinds of emotions, which in turn depends on having some skills in working with emotions. There are many such skills, beginning with the fundamental ability to recognize and distinguish between emotions. As with analytical skill in working with concepts—indeed, because it *is* (in part) an analytical skill in working with concepts—the finer the distinctions one can draw between various kinds of emotions, the more facility one has in working with emotions (Barrett, 2017, pp. 179–186). Another key skill is the ability to observe one's own emotions (and their connected thoughts) carefully but nonjudgmentally: that is, to be consciously aware of feeling emotions in the body as they are happening without being so fully identified with or immersed in them that nothing "outside" of the emotion is discernable.[18] A related skill is the ability to compassionately allow and "hold" emotions, to provide a supportive "container" in which one can experience emotions without reacting *automatically* in some way, either internally or externally. This skill is especially important for being able to tolerate difficult emotions without becoming either so distressed that one shuts down and pushes them away, or so drawn in that one needlessly fuels their perpetuation. Together these skills of distinguishing, observing, and holding emotions provide the foundation for the skill of investigating the source and nature and composition of one's emotions, allowing one to explore the networks of sensations, perceptions, and beliefs incorporated in the emotions and to learn from the information these networks carry.

The facilities of holding, understanding, and learning from emotions in turn strengthen the ability to work with emotions, providing more options for purposefully and judiciously cultivating, allowing, and/or transforming them. Emotions can be intentionally cultivated by observing and identifying which circumstances, perceptions, thoughts, bodily states, and activities tend to encourage which kinds

18. While not unique to Buddhism, this skill is central to Buddhist practice. In English it has become standard to refer to it as "mindfulness," which can unfortunately suggest that the skill is simply a form of thinking (using the "mind"). In fact, it is a more embodied practice that involves observing both thoughts and emotions without being fully identified with or immersed in them.

of emotions in oneself and then actively pursuing whatever fosters the desired emotions. In other words, one can nurture the emotions one would like to feel by nurturing the elements that make them up.[19] As just a simple example: if one has learned that certain kinds of music encourage certain kinds of moods in oneself, one can cultivate, say, feelings of joy or nostalgia or motivation by careful choice of soundtrack. Likewise, emotions can be intentionally transformed by altering the environmental and bodily contexts influencing them, perhaps as simply as by eating, sleeping, exercising, or going outside (Barrett, 2017, pp. 176–179), or in more complex ways by, for example, seeking out different friends. Emotions can also be transformed by (among other strategies) engaging in conscious and compassionate inner dialogue with them, questioning and shifting the perspectives and beliefs embedded within them, and recategorizing them. For instance, one might reinterpret "anxiety" as "anticipation" or "determination" in order to shift the experience of the associated sensations, as one karate teacher encourages when he tells his nervous young students to "make your butterflies fly in formation" (Barrett, 2017, p. 189). Again, this is not to say that people can construct, deconstruct, and reconstruct emotions just exactly as they please, and certainly not instantaneously. But in fact almost every adult already does this to a significant extent. More importantly, most of us could probably learn how to do it better.

All of the emotional work necessary to develop and sustain environmental virtues requires such facility in working with emotions. The virtues discussed in the previous section—responsibility, humility, forgiveness, creativity, open-mindedness, and participatory virtues—all require establishing a consistent emotional investment in specific values, nurturing emotions necessary to support those

19. This is precisely what Aristotle advocates by recommending the habituation of emotions (Kosman, 1980; Sherman, 1995, 1999b).

values, and identifying and transforming emotions that can undermine support for those values. Absent some level of emotional skill, none of these virtues would be possible.

CONCLUSION

As dispositions of character, virtues are patterns of cognitively/emotionally assessing and responding to values and circumstances in ways that are at least arguably admirable as excellent. Too often, however, arguments for environmental virtues make the case for a necessary virtue without much attention to the emotional aspects involved in developing and sustaining that virtue. My argument in this chapter has been that virtues of all kinds must be cultivated at least in part through cultivating particular ways of emotionally engaging with the world. In turn, cultivating particular ways of emotionally engaging with the world requires facility in working with emotions. Of course, this is true for environmental virtues as well. Indeed, insofar as environmental virtues stand against hegemonic norms and systems in some contexts, their development may require even more vigorous cultivation and stronger emotional skills than other virtues do.

One conclusion that follows from this argument is that if we want to promote environmental virtues of any kind, we need structures that will support their cultivation, including the development of emotional facility in general. Possible structures certainly include formal education at all levels but also include workshops, training programs, guided practices, counseling, support groups, and more, whether in public or private contexts. Models to emulate, mentors to provide guidance and feedback, and peers to share experiences with are all important. Above all, we need structures that will support people in *practicing* the skills of working with emotions and developing virtues, since, as Aristotle noted long ago, praxis is key to developing moral life (Kosman, 1980, p. 115). But notice as well that we have now returned to a point from the beginning: creating a more sustainable future requires work on both individual and collective/institutional levels. Individual virtuous actions are necessary, but they must be collectively organized in order to challenge the systemic forces driving environmental degradation *and* they must be fostered by social institutions that nurture their development in the first place.

REFERENCES

Abbate, C. 2014. "Virtues and animals: A minimally decent ethic for practical living in a non-ideal world." *Journal of Agricultural and Environmental Ethics, 27,* 909–929.

Bannon, B. 2017. "Being a friend to nature: Environmental virtues and ethical ideals." *Ethics, Policy & Environment, 20*(1), 44–58.

Barrett, L.F. 2017. *How Emotions Are Made: The Secret Life of the Brain.* Boston: Houghton Mifflin Harcourt.

Bendik-Keymer, J. 2010. "Species extinction and the vice of thoughtlessness: The importance of spiritual exercises for learning virtue." *Journal of Agricultural & Environmental Ethics,* 23(1, 2), 61–83.

Bendik-Keymer, J. 2012. "The sixth mass extinction is caused by us." In *Ethical Adaptation to Climate Change: Human Virtues of the Future,* edited by J. Bendik-Keymer and A. Thompson, 263–280. Cambridge, MA: The MIT Press.

Cafaro, P. 2005. "Gluttony, arrogance, greed, and apathy: An exploration of environmental vice." In *Environmental Virtue Ethics,* edited by R. Sandler and P. Cafaro, 135–158. Lanham, MD: Rowman & Littlefield.

Cafaro, P. 2010. "Patriotism as an environmental virtue." *Journal of Agricultural and Environmental Ethics,* 23(1, 2), 185–206.

Calhoun, C. 1984. "Cognitive emotions?" In *What Is an Emotion? Classic Readings in Philosophical Psychology,* edited by C. Calhoun and R.C. Solomon, 327–342. New York: Oxford University Press.

Clowney, D. 2013. "Biophilia as an environmental virtue." *Journal of Agricultural & Environmental Ethics, 26*(5), 999–1014.

Clowney, D. 2014. "Collective environmental virtue." *Environmental Values,* 23(3), 315–333.

Coeckelbergh, M. 2015. *Environmental Skill: Motivation, Knowledge, and the Possibility of a Non-Romantic Environmental Ethics.* New York: Routledge.

Cooper, J. 1999. *Reason and Emotion: Essays on Ancient Moral Psychology and Ethical Theory*. Princeton, NJ: Princeton University Press.

de Sousa, R. 1980. "The rationality of emotions." In *Explaining Emotions*, edited by A. O. Rorty, 127–152. Berkeley: University of California Press.

Dixon, M.H. 2010. "Environmental virtue ethics: Core concepts and values." In *Managing Environmental Justice*, edited by D. Pavlich, 3–22. Amsterdam: Rodopi B.V.

Ferkany, M. 2011. "Mercy as an environmental virtue." *Environmental Values*, *20*(2), 265–283.

Ferkany, M., and K.P. Whyte. 2012. "The importance of participatory virtues in the future of environmental education." *Journal of Agricultural & Environmental Ethics*, 25(3), 419–434.

Frasz, G. 2005. "Benevolence as an environmental virtue." In *Environmental Virtue Ethics*, edited by R.D. Sandler and P. Cafaro, 121–134. Lanham, MD: Rowman & Littlefield.

Fredericks, R. 2014. "Courage as an environmental virtue." *Environmental Ethics*, *36*(3), 339–355.

Gambrel, J.C., and P. Cafaro. 2010. "The virtue of simplicity." *Journal of Agricultural & Environmental Ethics*, 23(1,2), 85–108.

Global Footprint Network. n.d. "Global Footprint Network." Accessed May 27, 2019. https://data.footprintnetwork.org/#/sustainableDevelopment?cn=all&type=earth&yr=2016.

Hall, C. 2016. "Framing and nudging for a greener future." In *The Oxford Handbook of Environmental Political Theory*, edited by T. Gabrielson, C. Hall, J.M. Meyer, and D. Schlosberg, 593–607. Oxford: Oxford University Press.

Hall, C. 2019. "Caring to be green: The importance of love for environmental integrity." In *The Routledge Handbook of Love in Philosophy*, edited by A.M. Martin, 215–226. New York: Routledge.

Hannis, M. 2015. "The virtues of acknowledged ecological dependence: Sustainability, autonomy and human flourishing." *Environmental Values*, *24*(2), 145–164.

Hulme, M. 2014. "Climate change and virtue: An apologetic." *Humanities*, 3(3), 299–312.

Hursthouse, R. 2011. "Virtue ethics and the treatment of animals." In *The Oxford Handbook of Animal Ethics*, edited by T.L. Beauchamp and R.G. Frey, 119–143. Oxford: Oxford University Press.

Hursthouse, R. 2007. "Environmental virtue ethics." In *Working Virtue : Virtue Ethics and Contemporary Moral Problems*, edited by P.J. Ivanhoe and R.L. Walker, 155–171. Oxford: Clarendon Press.

Jacobs, S. 2016. "Bad news: Low-carbon air travel isn't very likely." *Grist* (blog). March 4, 2016. http://grist.org/business-technology/bad-news-low-carbon-air-travel-isnt-very-likely/.

Kawall, J. 2003. "Reverence for life as a viable environmental virtue." *Environmental Ethics*, 25(4), 339–358.

Kawall, J. 2018. "Environmental virtue ethics." In *The Oxford Handbook of Virtue*, edited by N.E. Snow, 659–679. Oxford: Oxford University Press.

Kosman, L. A. 1980. "Being properly affected: Virtues and feelings in Aristotle's ethics." In *Essays on Aristotle's Ethics*, edited by A.O. Rorty, 103–116. Berkeley: University of California Press.

Lenzi, D. 2017. "Relativism, ambiguity and the environmental virtues." *Environmental Values*, 26(1), 91–109.

Levin, K. 2018. "8 things you need to know about the IPCC 1.5°C Report." *World Resources Institute* (blog). October 7, 2018. https://www.wri.org/blog/2018/10/8-things-you-need-know-about-ipcc-15-c-report.

Loder, R.E. 2011. "Gratitude and the environment: Toward individual and collective ecological virtue." *Journal Jurisprudence*, *10*, 383–435.

Mason, K. 2014. "Becoming citizen green: Prefigurative politics, autonomous geographies, and hoping against hope." *Environmental Politics*, 23(1), 140–158.

Norlock, K.J. 2010. "Forgivingness, pessimism, and environmental citizenship." *Journal of Agricultural and Environmental Ethics*, 23(1,2), 29–42.

Nussbaum, M.C. 1994. *The Therapy of Desire: Theory and Practice in Hellenistic Ethics*. Princeton, NJ: Princeton University Press.

Nussbaum, M.C. 2001. *Upheavals of Thought: The Intelligence of Emotions*. Cambridge: Cambridge University Press.

Ockwell, D., L. Whitmarsh, and S. O'Neill. 2009. "Reorienting climate change communication for effective mitigation: Forcing people to be green or fostering grass-roots engagement?" *Science Communication*, 30(3), 305–327.

Pianalto, M. 2013. "Humility and environmental virtue ethics." In *Virtues in Action: New Essays in Applied Virtue Ethics*, edited by M. Austin, 132–149. New York: Palgrave Macmillan.

Plumer, B. 2016. "Solar planes aren't the green future of air travel. But here's what could be." *Vox*. May 6, 2016. https://www.vox.com/2016/5/6/11569202/aviation-emissions-solar-plane.

Sherman, N. 1995. "The role of emotions in Aristotelian virtue." *Proceedings of the Boston Area Colloquium in Ancient Philosophy*, *9*, 1–33.

Sherman, N. 1999a. "Taking responsibility for our emotions." *Social Philosophy & Policy*, *16*(2), 294–323.

Sherman, N. 1999b. "The habituation of character." In *Aristotle's Ethics: Critical Essays*, edited by N. Sherman, 231–260. Lanham, MD: Rowman & Littlefield.

Sherman, N. 2000. "Wise emotions." In *Understanding Wisdom: Sources, Science, & Society*, edited by W.S. Brown, 319–337. Laws of Life Symposia Series; Vol. 3. West Conshohocken, PA: Templeton Foundation Press.

Solomon, R. 2003. *Not Passion's Slave: Emotions and Choice*. Oxford: Oxford University Press.

Thiele, L.P. 2011. *Indra's Net and the Midas Touch: Living Sustainably in a Connected World.* Cambridge, MA: MIT Press.

Thompson, A. 2010. "Radical hope for living well in a warmer world." *Journal of Agricultural & Environmental Ethics,* 23(1,2), 43–59.

Thompson, A. 2012. "The virtue of responsibility for the global climate." In *Ethical Adaptation to Climate Change: Human Virtues of the Future,* edited by J. Bendik-Keymer and A. Thompson, 203–222. Cambridge, MA: The MIT Press.

Treanor, B. 2010. "Environmentalism and public virtue." *Journal of Agricultural & Environmental Ethics,* 23(1,2), 9–28.

Vogel, S. 2012. "Alienation and the commons." In *Ethical Adaptation to Climate Change: Human Virtues of the Future,* edited by J. Bendik-Keymer and A. Thompson, 299–315. Cambridge, MA: The MIT Press.

Wood, D. 2018. "Global Air Passenger Traffic Soared in 2017." TravelPulse.com, February 2, 2018. https://www.travelpulse.com/news/airlines/global-air-passenger-traffic-soared-in-2017.html

World Commission on Environment and Development. 1987. *Our Common Future.* Oxford: Oxford University Press.

PART II

VIRTUES, SUSTAINABILITY, AND WORLDVIEWS

Chapter 5

Jain Dharma as a Virtue Ethics for Sustainability

PANKAJ JAIN

INTRODUCTION

In her presidential address at the American Political Science Association's annual meeting, Susanne Rudolph mentioned (2005):

> Early in our research in India, Lloyd Rudolph and I coined the phrase "imperialism of categories." It was meant to designate the academic practice of imposing concepts on the other—the export of concepts as part of a hegemonic relationship. Categories crafted in a dominant socio-cultural environment are exported to a subordinate one. The imperialism of categories entails an unself-conscious parochialism of categories: scholars from a dominant culture, sometimes called the center, travel to a distant and lesser place, sometimes called the periphery, where they apply "universal" concepts. The trouble is that the concepts have been fashioned out of the center. The East is fatalist, says Max Weber; the West, agentic. The non-West conveys status

Pankaj Jain, *Jain Dharma as a Virtue Ethics for Sustainability* In: *The Virtues of Sustainability*.
Edited by: Jason Kawall, Oxford University Press (2021). © Oxford University Press.
DOI: 10.1093/oso/9780190919818.003.0005

> by birth, says Talcott Parsons; the West, by achievement. The non-West is childlike, says John Stuart Mill; the West, mature. Dominant peoples use ideal types and stereotypes to control the dominated by ranking and creating cultural social registers.

To avoid such "imperialism of categories" and also following Gerald Larson (2004) and McKim Marriott (1990), in this chapter, I begin by noting that terms such as philosophy, religion, ethics, sustainability, and environmentalism are all, obviously, English terms and etymologically based on Greco-Roman intellectual heritage and/or Judeo-Christian religious heritage. In much of the non-Western world, such as in Asia and Africa, almost all such ideas remain intertwined. For instance, in virtually all the texts in Indian history and contexts from the contemporary communities in India, it can be a useless exercise to explore where philosophy ends and where religion begins or vice versa. In these descriptions and prescriptions for thinking and living in the world, philosophy and religion overlap, and sustainability and environmentalism remain an undercurrent, adding a third layer to this combination of philosophy and religion.

Among the ancient religions of the world, Jainism and Buddhism are usually categorized as nontheistic religions. In this chapter, I will explore how karma theory and virtues such as nonviolence guide Jains in their actions. Rather than investigating the "virtue ethics" of dharma for sustainability, perhaps we can begin by looking at the Jain way of life. This unique tradition brings to our attention such unique examples as some monks having renounced even their clothes. Similarly, most Jain laypeople have been vegetarian for more than 2,600 years, millennia before it became a trend in Euro-America. The last great teacher in the Jain tradition was Mahavira, who was the contemporary of the Buddha (living around sixth century BCE). The combined influence of Buddhism and Jainism has profoundly shaped Indian society, most evidently in the practice of vegetarianism that millions of Indians continue to practice today in India and now across the world in the diaspora. While Buddhism also encourages nonviolence and vegetarianism, Jainism put the ultimate emphasis on this ideal and defines the religion or philosophy itself as nonviolence with its great mantra *Ahimsa Paramo Dharma*, literally "nonviolence is the supreme virtue, ethic, and religion."

In America, nonviolence is perhaps most closely associated with Dr. Martin Luther King, who based his civil rights movement on this principle. Dr. King and a few other African American leaders had visited the home of Gandhi in India and brought the Gandhian ideal of nonviolence to America (Lakshmi, 2009). Gandhi, in turn, was greatly influenced by Jainism. So some scholars have referred to Jainism with the three M's who emphasized the principle of non-violence: Mahavira (the last celebrated teacher contemporary of the Buddha), Mahatma (Gandhi), and Martin (Luther King) (Helton, 2007; King, 1999; Nojeim, 2004). Thus, at least in two ways, Jainism's nonviolence remains relevant in the contemporary United States: the enduring influence of Dr. King's method of nonviolent resistance and increasing awareness for vegetarianism. In the last decade, there have been several reports from the United Nations (see FAO, n.d.) and other agencies connecting meat consumption with climate change; for instance, factory farming is now widely seen as one of the extreme examples of cruelty against animals and, in turn, toward our planet.

THE INFLUENCE OF JAIN VIRTUE ETHICS ON INDIA'S SUSTAINABILITY AND ENVIRONMENTAL ISSUES

Prima facie, one can be appalled by the large-scale destruction and pollution of India's natural resources at present. The assault on land is visible everywhere. The forest areas are under tremendous pressure, especially as villagers continue to look for their daily firewood and fodder in their natural environment. The problem of encroachment of such open areas is also a severe threat. Industrial pollution in India is akin to what happened to American cities such as Pittsburgh in the 1940s (Davidson, 1979). One can also notice the unplanned urbanization leading to issues of waste disposal, for instance. There are several large-scale projects with significant impacts, such as large dams that do not necessarily rehabilitate and relocate the displaced indigenous people from such sites. The pollution in Indian rivers can remind one of the American rivers such as Ohio's Cuyahoga in the 1960s (Rotman, 2010). And finally, the Indian population continues to expand and is going to surpass that of China by 2028, according to the World Population Prospects Report by the United Nations.

However, to present an alternative picture of India's environment, we can look at the Greendex survey conducted for several years by the National Geographic between 2008 and 2014 (see, for example, Greendex, 2009). In these surveys, India consistently appears as the "greenest" country, greener than the United States and other Western countries as well as greener than China and Brazil, other developing countries. The criterion that kept India greener than other countries is the national carbon footprint of Indians that remains lower than their counterparts elsewhere, based on broad parameters such as consumption in terms of housing needs, transportation needs, food, and other consumer goods (for further discussion of the importance of ecological footprints, see Vanderheiden, Chapter 9, this volume). In each of these four categories, a large number of Indians continue to keep their carbon footprint lower than people elsewhere. This frugal lifestyle of Indians used to be critiqued by economists (sometimes disparagingly) by labeling it as the "Hindu rate of economic growth." These economists lamented that Indians consumed fewer natural resources than their counterparts elsewhere, not just because of lower per capita income levels but also because of the influence from philosophical and cultural traditions such as Hinduism. Be that as it may, one can undoubtedly note that the dietary habits of many Indians were influenced more than 2,000 years ago when the great leaders of Buddhism and Jainism, the Buddha and Mahavira respectively, emphasized nonviolence toward animals to be one of the supreme virtues. Over time, most Indians stopped sacrificing animals in their rituals, and today they depend on rice, wheat, lentils, fruits, and vegetables for their staple diet (although some of them might choose a chicken-based meal occasionally). So, while economists critiqued the "Hindu rate of growth," environmentalists should probably celebrate the "Hindu-Jain way of life" that keeps India's carbon footprint lower than that of other countries. In the rest of this chapter, I will explore the rationale for Jain virtue ethics from the perspective of sustainability.

Jain cosmology divides all the beings in the universe into two categories, living (*jiva*) and nonliving (*ajiva*) and space into two categories, the living abode (*loka*), and nonliving abode (*aloka*). In this cosmology, in addition to humans, animals, birds, insects, reptiles, and even earth, fire, air, and water are considered to be *jiva*, that is, with distinct and unique souls. The prominent nonliving beings, in addition to the medium of motion, medium of rest, time, and space, are the karma particles that pollute the souls of most of us because of the violence that we commit, intentionally or unintentionally, as we live in the world. Since the complete cessation of violence is not possible for any living being, out of one-, two-, three-, four-, and five-sensed beings, Jainism encouraged humans to consume only one-sensed beings (Chapple, 2002, 128). These are earth, air, fire, water, and plants that all have only one sense of touch. Five-sensed beings include mammals (including humans) and birds.[1] Jainism's rationale

1. According to Jainism, two-sensed beings are worms, leeches, oysters, and snails with lifespan of 12 years, three-sensed beings are insects and spiders with lifespan of 49 days, and four-sensed beings are butterflies, flies, and bees with lifespan of six months.

for nonviolence equates humans with other mammals and birds, so for a human to kill any animal or bird would accrue the same karma as killing a fellow human being.

Further, for human living, even one-sensed beings such as earth, water, air, fire, and plants should be consumed with the utmost care. Among plants, vegetables such as carrots, radishes, onions, and garlic are discouraged because destroying a stem or root is killing the entire plant. Similarly, only filtered water is to be consumed to avoid any violence to microbes in the water. Some Jain monks and nuns cover their mouths as they speak and breathe to prevent any harm to microbes in the air. Many devout Jains continue to avoid eating after sunset to minimize the consumption of energy that comes by burning fossil fuel. Similarly, Jain monks and nuns avoid riding vehicles as well. Based on similar concerns for one-sensed beings, many Jains today are actively trying to preserve wildlands, and keep waterways and air unpolluted, in particular, and address climate change, in general.

In addition to *ahimsā* (nonviolence), *aparigraha* (nonaccumulation of materials) is the next important ideal that has inspired Jain virtue ethics in its long history. Just as nonviolence finds its most clear application in vegetarianism, *aparigraha* finds its application in the Jain insistence in minimizing worldly possessions, as explained in the second book of one of the earliest Jain texts, *Ācārāṅga Sūtra* (Dundas, 2003, p. 159). In an article for the *Washington Post* (December 14, 2006), Diane Eck quoted Gandhi's words, "The earth has enough for one's need but not for one's greed." Gandhi practiced what he preached by limiting his possessions and even his necessities.

Several Jains have also donated large sums of their hard-earned wealth for philanthropic causes such as education, healthcare, disaster reliefs, orphanage, and animal shelters, to name a few. Lay Jains have continued to support and revere their monks and nuns, as their ultimate role models for *aparigraha*. Indeed, some monks go to the extreme of renouncing every last piece of cloth and utensils for eating their food. They walk entirely naked and eat a small meal with their bare hands because even collecting food or wearing clothes may turn into attachment and some violence toward natural resources. Such monks continue to inspire thousands of Jains to carry the ideal of *aparigraha* in high esteem and a goal to achieve as they hope to progress in their spiritual journey. Similar to nonviolence, nonaccumulation and nonpossession, therefore, also serve as high virtues for Jains for the sustainability of our planetary resources. These Jain virtues compare well with the conscientiousness that Vanderheiden discusses in his chapter in this volume (Chapter 9). That is, nonviolence and nonpossession may seem to require us to pay attention to our actions and way of life carefully—noticing small impacts, and so on—in ways that might have some similarities (and, of course, some differences) to Vanderheiden's conscientiousness.

The third crucial Jain virtue, *anekāntavāda,* means the doctrine of pluralism or "many pointedness" (Dundas, 2003, p. 229). This virtue continues to inspire Jains to practice intellectual nonviolence by trying to keep an open mind toward different philosophical and religious traditions. The parable of an elephant and six blind men explains this virtue well. As these men touch different parts of the elephant, they equate the elephant with a snake (the man touching its tail), a wall (the man touching its belly), a spear (the man touching its tusk), and so on. However, the holistic definition of an elephant is possible only by combining all these partial definitions. *Anekāntavāda* similarly has inspired Jains to learn and preserve their rival ideologies of Hinduism and Buddhism for centuries. Because of this open mind, Jains did not hesitate to interact with Islamic emperors and even partially succeeded in making them adopt nonviolence as the state policy occasionally. The virtue of *anekāntavāda* also applies to sustainability by inspiring Jains to consider the natural resources also from nonanthropocentric perspectives. For instance, in India, many have observed that even the national highways are open for animals, unlike the American national highways, where even humans are not allowed to "loiter." Why should humans monopolize the roads or any other planetary resource? *Anekāntavāda* encourages that we accept even the animals' perspectives toward the highways, water, or other natural resources.

Chapter 7 in this volume by Laura M. Hartman discusses "cooperativeness" as a virtue of sustainability. Perhaps Jainism's *anekāntavāda* can help to achieve cooperation and sustainability. For example, insofar as it requires thinking about the perspectives of others, it seems it could be conducive to encouraging social sustainability, and also alliances for environmental sustainability. Nonviolence and pluralism might be very important for sustainability in terms of allowing diverse communities to tolerate and respect each other, and to work together for sustainability. These Jain values might be beneficial in social movements for social and environmental sustainability.

JAIN ROLE MODELS AND THEIR INSPIRATION FOR VIRTUE ETHICS

As aforementioned, Mahavira (~600 BCE), the great hero, was the 24th and the last great teacher of Jainism and was a contemporary of the Buddha. Many of his statements and actions concerning compassion for trees and animals continue to inspire many Jains to be protectors of animals and nature in general even today. For instance, one of the prominent Jain sacred texts *Ācārāṅga Sutra* records his statements such as "trees are inherently valuable" (2.4.2.11–12) and "vegetation has a life just as humans" (1.1.5.9–11). Just before Mahavira is about to pass away, he sent one of his foremost disciples, Indrabhuti, to stop the slaughter of cattle.

In Jain cosmology, there are 24 such great teachers in every era. Before Mahavira, the 22nd teacher Neminatha and the 23rd teacher

Parshvanatha both renounced their royal families and wealth as soon as they encountered violence against animals and reptiles, respectively. Other Jain sacred texts such as *Tattvārtha Sutra* and *Bhavāni Sutra* have statements such as "The souls render service to one another." Another verse from *Dashavaikālika Sutra* (1.2–5) highlights the behavior of bees of collecting only minimal nectar from each flower instead of damaging or destroying the flower in the process. Humans should also preserve their natural resources, and they should consume them minimally for their existence. Another text, *Adipurāna*, states, "Forests are like saints and trees should be planted for positive karma."

Such examples and role models have continued to inspire Jain householders to get involved in relatively less violent professions such as trading, banking, administrative activities, traditional medicine, arts, crafts, publishing, and education. According to an ancient Jain text, *Bhogopabhog Vrata*, one would rarely, if ever, find a Jain householder involved in relatively more violent professions such as trading of charcoal, alcohol, slaves, weapons, poisons, sugar canes, firearms, or timber. Jains have stayed away from exploiting animals or constructing lakes, as such activities also involve causing great violence to the environment.

CRITIQUING CORT'S CRITICISM

Despite the many promising features of Jain virtue ethics for achieving and maintaining sustainability that we have seen, it is not without its critics. John Cort challenges the project of connecting Jainism with ecology or environmental ethics, charging that "The historical record of the Jains is, on the whole, not a positive one" (2002, p. 68). Without giving us specific details, Cort chastises Jains to be "honest" about admitting their environmentally destructive habits in these lines. Cort states that Jainism is based on a body-spirit dualism when he writes:

> [The Jain] ideology is not very conducive to the development of an environmental ethic. At its heart is the goal of permanent separation of the soul from all matter. In such a dualist ideology, positive environmental results are largely incidental. (2002, p. 70)

The spirit (*jiva*) is indeed separate from the matter in the Jain universe (*loka*). However, as Cort mentions later in his chapter, a unique and distinct *jiva* is present in the earth-bodies, water-bodies, fire-bodies, and air-bodies, in addition to all humans, animals, birds, plants, gods, and demons. The very fact that Jainism holds that almost every particle is alive with a *jiva* (soul or spirit) makes Jainism, in my view, an exceptional ideology. It is because of this recognition of *jiva* in plants, animals, earth, water, air, and fire that most Jains continue to be vegetarians, thus keeping their carbon footprints considerably lower than their Euro-American counterparts. This Jain zeal for avoiding meat as a staple diet has also influenced their Hindu counterparts over the millennia and can be seen as a key factor contributing to India continuing to have the lowest carbon footprint among significant economies such as China, Brazil, and other Euro-American countries, as shown by the Greendex surveys produced by National Geographic (Greendex, 2009).

Further, Cort notes that the absence of a creator God in Jainism and the presence of *jiva* universally in all beings make Jainism an exceptional tradition. Not the fear of reprisal from God, but the *jiva* equality of all lives in the universe is the emphasis in Jainism. Cort is right in advocating that Jains should extend their nonviolence more proactively toward earth, air, water, and fire, in addition to their concern for animals (2002, p. 75). This greater nonviolence might mean, for example, hesitating to build dams or resisting mountaintop removal mining or strip mining.

While discussing environmental justice, Cort cites the example of reforestation efforts of the Shatrunjaya hills (2002, p. 79). The reforestation efforts, according to Cort, did not take into consideration the needs of local herders who used to graze their cattle and goats on those mountains. According to Cort, even a laudable environmental effort by Jains is of little value if it hurts the local people. He writes,

> [H]ere we see an effort to improve the environment but at the expense of those who are the poorest and most dependent upon the resources, and who have not been involved in either the decision-making or implementation processes. The laudable ecological effort to reforest the mountain has dispossessed the poorer herders of an important means of livelihood. (2002, p. 79)

This point has significant climatic ramifications. With dozens of reports by the United Nations and scientific agencies, we know that thousands of acres of forests have been destroyed by ranchers who replace forests for animal farming (FAO, n.d.; Koneswaran and Nierenberg, 2008). Although Jains' reforestation decisions need to incorporate local needs, the Jains' respect for forests and the global destruction of forests by the powerful lobbies of the meat industry need to be seriously studied by the scholars of religion and ecology. Jainism has much to contribute to both sides of this issue. Unlike many American ranchers who have little regard for local ecological phenomena, Jains are now trying to reforest the barren lands, for instance, at the Shatrunjaya hills, as noted by Cort. However, Cort and scholars of environmental ethics need to direct their attention to activities that are actively destroying the lungs of our planet. The massive consumption of meat, a staple diet of most Americans, is a product of animal farming and is one of the primary reasons for Amazonian forest destruction and climate change. As some, such as

Ramachandra Guha (2006, p. 225), have noted, the real "population problem" is in America with its unsustainable consumption habits. In addition to humans, the population of American cattle is also primarily raised unsustainably by ranchers at the enormous cost to the forests, such as the Amazonian rainforest.

Cort continues his critique by including Jains among the global North, whose massive consumption is significantly endangering our planet's resources. This stereotypical image of wealthy Jains ignores the fact that Jains have been living a much simpler and sustainable life compared to their counterparts in industrialized Western economies. As already mentioned, food is the most significant difference that keeps Jains' carbon footprint much lower than their counterparts in Euro-American nations. Other values such as *aparigraha* (nonaccumulation of resources) also have historically inspired Jains to be at the forefront of donating much of their wealth for charitable causes such as disaster relief, building sanctuaries for sick and diseased animals, rest houses for travelers, and educational and healthcare institutions across India, as also noted by Padmanabh Jaini (2002, p. 153).

Elsewhere in his chapter, while chiding Jains for their "narrow" nonviolence, Cort holds that Jains need to expand the domain of their nonviolence against global wars and military violence. However, Cort fails to mention prominent Jain monks such as Sushil Muni, Mahāprajña, and Tulsi, who tirelessly worked to stop all military and terrorism-related violence across the world (Gandhi, 2015). Mahatma Gandhi remains the icon for nonviolence around the globe, and the influence of Jainism on him is well known among the scholars and practitioners of Jainism.

Rather than searching for Western concepts such as "Deep Ecology" in Jainism, I believe it is appropriate to discuss an "Ahimsak Ecology" that is rooted in the taxonomy based on concepts such as *jiva, ajiva,* and *loka*. Building on Cort's concern for local ecology and bioregionalism, it is time to develop and discuss localized, and regionally distinct, categories in the humanities and social sciences, rather than imposing universalizing categories and terms originating in Greek, Roman, Jewish, or Christian thought traditions. Cort hints at this but does not develop the idea in his chapter when he says, "Understanding of and reactions to the environment are always *local*" (2002, p. 67). The rest of this chapter will examine prominent and contemporary Jain individuals and how their virtues lead to wide-ranging, sustainable behaviors.

GANDHI AND OTHER EXAMPLES OF JAIN VIRTUE ETHICS FOR SUSTAINABILITY

Mahatma Gandhi remains the most prominent role model that practiced and preached the virtue ethics of sustainability based on Jainism (and other philosophical traditions). Vinay Lal (2000) describes Gandhi as "too deep" even for Deep Ecology. According to Lal, by leading a virtuous life, Gandhi serves as a role model of environmentalism that goes far beyond what Deep Ecology advocates. Although Gandhi left for South Africa in 1891, he continued to correspond with a Jain guru, Shrimad Rajchandra, for several years. He adopted several important ideas into his movement against the British colonial and racist policies against Indians in South Africa and later in India. Some of these virtues were the truth (*satya*), nonviolence (*ahimsa*), nonaccumulation (*aparigraha*), celibacy (*brahmacharya*), self-reliance (*purushartha*), and fasting (*upavasa*). Gandhi's life emerged as an ecological treatise, a living ecosystem as manifested by his lifelong practices such as his small meals of nuts and fruits, his daily ablutions and bodily practices, his periodic observances of silence, his morning walks, his spinning wheel, and his avoidance of waste. Today many moralists, nonviolent activists, feminists, journalists, social reformers, trade union leaders, peasants, prohibitionists, nature-cure lovers, renouncers, and environmentalists continue to take their inspiration from Gandhi's life.

Long before Gandhi, emperor Aśoka erected dozens of edicts across India to proclaim nonviolence in his empire by maintaining sacred groves and prohibiting hunting and violence (Dhammika and Aśoka, 1993). Although Aśoka adopted Buddhism, his grandfather Chandragupta Maurya (Encyclopedia Britannica, 2020) and grandson Samprati (Dundas, 2003, p. 115) had adopted Jainism, according to Jainism's historical sources. Similarly, an emperor in Gujarat, Siddharaja, and his successor, Kumarapala, promoted Jainism and nonviolence in the medieval period in Indian history. The Islamic emperors such as Muhammad Tughluq, Akbar, and Jahangir also prohibited hunting and meat consumption for some special days periodically under the influence of Jain monks who visited their courts.

JAINS, SOCIAL ACTION, AND SUSTAINABILITY TODAY

As Flugel (2005) notes, Jains have been involved in social activism in their long history. Recent prominent Jain leaders and spokespersons demonstrate the Jain adherence to nonviolence, such as the late Jain monks Sushil Muni and Acharya Tulsi, who opposed geopolitical conflicts, terrorism, and other kinds of socio-political violence. As Cort (2002) notes, hundreds of thousands of Jains perform a ritual called *Pratikraman* every day in which they ask for forgiveness from 8.4 million species,[2] including all of the one-sensed, two-sensed,

2. Although the total number of species according to the latest scientific research is approximately 8.7 million (±1.3 million) according to Mora et al. (2011), the number of species in all the ancient Indian texts is mentioned to be 8.4 million. If biologists tell us that there are over 10 million species, presumably Jains would respect and seek forgiveness from all of them.

three-sensed, four-sensed, and five-sensed beings, for any intentional or unintentional violence committed in action, speech, or thought. This ritual, if understood sincerely, can help to remind and transform some of those Jains that are involved in industries that are polluting and damaging local natural resources. Many Jains continue to support and fund hundreds of shelters and hospitals for animals and birds across India and North America (Lodrick, 1981). The Bird Hospital in New Delhi remains one of the most celebrated such efforts (Andersen, 2019). From its establishment in 1956, Jains continue to support and manage this avian veterinary hospital that treats thousands of sick and injured birds each year.

Khichan, a small village in the desert of the Indian state Rajasthan, has similarly caught the attention of several birdwatchers as the local Jains there welcome and feed thousands of Demoiselle Cranes that visit this village from across Asia. Another example is from Goa in India, where the largest wildlife sanctuary was named after Mahavira, the last great teacher of Jainism, and the local Jains there contributed a large sum of money for this cause (Jain, 2011). Jains continue to promote vegetarianism and push for the ban on meat consumption across the world. According to several recent news reports, a number of Jain temples such as those in Byculla, Dahanu, Nashik, Shahpur (Maharashtra, India), Dilwara (Rajasthan, India), Kolaththur (Tamil Nadu, India), Bareilly (Uttar Pradesh, India), Franklin (New Jersey, USA), and Milpitas (California, USA) have adopted solar energy to reduce their carbon footprint. Other Jain temples, such as those in Detroit, Michigan, and Dallas, Texas, have also taken other green measures as a reflection of the Jain virtue of nonviolence toward nature in general. Other Indian Jain temples such as Shatrunjaya Temple in Palitana, India, and Veerayatan, Bihar, have taken initiatives in reforestation in their surroundings. The Jain Center of Southern California holds ongoing events in support of animal rights. The Detroit Jain Temple works with the Sasha Farm, the Midwest's most significant animal sanctuary, housing over 200 rescued and aged animals near Ann Arbor, Michigan. Other American Jains are active in the Vegetarian Society of the District of Columbia and with another cattle sanctuary managed by the Voith Family of Angelica, New York (Vallely, 2002).

One of the most prominent examples of animal shelters is *Shree Sumati Jeev Raksha Kendra* in Rajasthan, India (Pavapuri, n.d.). A leading Jain group and other Jain donors manage this cattle sanctuary.[3] Its campus is over 7 million square feet and includes a Jain temple as well. This center started in 1998 with around 100 cattle and now houses more than 6,000 animals that are taken care of by a staff of 175 employees. Unlike North American cattle farms that are slaughtered for human consumption, the cattle at this center are managed according to Jain principles of feeding them only vegetarian food. These cattle are maintained in a sacred environment where Jain chants continuously play on loudspeakers. The center runs on biogas that is produced here by using the cow dung from the cattle at this center. The water needs of the center are fulfilled by rainwater harvesting at the center itself. Food for the animals is locally grown, and the milk from the cows is fed to their calves. The center also supports and feeds local birds and stray dogs. The center has also planted more than 100,000 trees in its vicinity, including some of the sacred trees of the Jain tradition.[4]

Scholars and practitioners need to study these phenomena in Jainism that have now become global in scope. For example, consider the contemporary Jain leaders who are taking the values of Jainism to the new frontier of veganism. Although the majority of Jain ascetics are vegetarians, we now have several activists and mendicants who

3. In 2017, North American Jains donated around $50,000 to such animal shelters across India and in North America (JAINA, n.d.).

4. In my (2011), I explore several such cattle sanctuaries across India.

embrace nonviolence at an even deeper level, through veganism. Pradeep Shah, founder of "Nature's Law and Diet System" in Mumbai, is one of many prominent Jain activists, who organizes talks by Jain ascetics on veganism. Jain ascetics such as Labdhi Sagar, Vaibhav Shree,[5] Viharsh Sagar, Aryika Chaitanyamati,[6] and Dev Nandi are spearheading the vegan movement inspired by the Jain principle of nonviolence. There are active websites such as VeganJains.com and JainVegans.org, which promote veganism among Jains and others. Several UK-based Jain groups are turning vegan. US-born Jains frequently organize vegan panels at JAINA conventions as well as their local temples. For a glimpse of such efforts, following is my translation of the Gujrati discourse by Labdhi Sagar in Mumbai. Some claims made by him may be extreme; I am sharing my verbatim translation from an excerpt of his recorded lecture on August 13, 2017:

> Today, we treat cattle like raw materials and exploit and torture them to the extreme for maximum output. In Jainism, we are forbidden to take anything from other beings without their consent. Our gurus advise us against such actions, and we accumulate a lot of negative karmas if we still go ahead with such actions. Such actions will lead to our premature death in the next lives, as explained by the Jain principle of karma, action, and consequence. Even the government does not share the exact statistics about the number of animals used in the dairy and meat industry. However, according to the Food Safety and Standards Authority of India (FSSAI), 75% of Indian milk is contaminated with chemicals such as urea, caustic soda, paints, detergent powder, sodium chloride, shampoo, etc. We should consume milk in minimal quantity, and the milk must be obtained from the cattle that are treated like family.

5. For a representative talk, see Shree (2018).
6. For a representative talk by Chaitanyamati, with English translation, see Fauna Police (2019).

CONCLUSION

Chapple's 2002 edited book, *Jainism and Ecology: Nonviolence in the Web of Life*, was the first significant attempt in the Western hemisphere to study Jainism from an environmental perspective. However, most of the epistemological and hermeneutic methodologies that are applied by scholars, such as Cort, on non-Western traditions such as Jainism, are rooted in Western categories. The Western categories must be broadened by incorporating Indic terms in the English language such as *prakriti*, *jiva*, and *ahimsa*. Western scholarship must also look at the living laboratories of the ecological, sustainable sensibilities in lives such as that of Gandhi, lives that defy all neat categories of being an environmentalist or a philosopher or an activist. The nonviolence-based inspiration from Mahavir (Swami) to Mahatma (Gandhi) has continued to Martin (Luther King) and several other contemporary leaders such as Sunderlal Bahuguna and Nelson Mandela and it is high time that scholars of religion and ecology embrace this Ahimsak Ecology and not just Deep Ecology. Although no religion can claim to have anticipated our current environmental situation, contemporary Jain vegans take their inspiration from Jainism's nonviolence and can be seen as a unique response to the ongoing sustainability crisis.

REFERENCES

Andersen, R. 2019. "A journey into the animal mind: What science can tell us about how other creatures experience the world." *The Atlantic*, March 2019. https://www.theatlantic.com/magazine/archive/2019/03/what-the-crow-knows/580726/

Chapple, C.K., ed. 2002. *Jainism and Ecology: Nonviolence in the Web of Life*. Cambridge, MA: Harvard University Press.

Cort, J.E. 2002. "Green Jainism? Notes and queries toward a possible Jain environment ethic." In *Jainism and Ecology: Nonviolence in the Web of Life*, edited by C.K. Chapple, 63–94. Cambridge, MA: Harvard University Press.

Davidson, C.I. 1979. "Air pollution in Pittsburgh: A historical perspective." *Journal of the Air Pollution Control Association, 29*(10), 1035–1041.

Dhammika, S., and Aśoka. 1993. *The Edicts of King Asoka: An English Rendering.* Kandy, Sri Lanka: Buddhist Publication Society.

Dundas, P. 2003. *The Jains.* London: Routledge.

Encyclopedia Britannica, eds. 2020. "Chandragupta." In *Encyclopedia Britannica.* Accessed on April 28, 2020. Encyclopedia Britannica, Inc. https://www.britannica.com/biography/Chandragupta

FAO (Food and Agriculture Organization of the United Nations). n.d. "Key facts and findings." Accessed April 20, 2020. http://www.fao.org/news/story/en/item/197623/icode/

Fauna Police. 2019. "Jain saint Chelna didi's (Aryika shree Chaitanyamati mataji) views". Facebook, June 22, 2019. https://www.facebook.com/FaunaPolice/posts/2986325688068226.

Flugel, P. 2005. "Review of *Jainism and Ecology: Non-Violence in the West of Life,* edited by Christopher Key Chapple." *Environmental Ethics, 27*(2), 201–206.

Gandhi, S. L. 2015. *Acharya Mahapragya: A Journey to Wisdom.* Noida, India: Harper Element.

Greendex. 2009. *Greendex 2009: Consumer Choice and The Environment—A Worldwide Tracking Survey.* National Geographic & GlobeScan. Accessed April 6, 2019. https://www.nationalgeographic.com/greendex/assets/GS_NGS_Full_Report_May09.pdf.

Guha, R. 2006. *How Much Should a Person Consume? Environmentalism in India and the United States.* Berkeley: University of California Press.

Helton, T. 2007. "Mahavira, Mohandas, and Martin: An ancient religion's influence on the United States civil rights movement." Unpublished paper.

Jain, P. 2011. "Jainism and environmental ethics." *Union Seminary Quarterly Review, 61*(1), 121–135.

JAINA (Federation of Jain Associations in North America). n.d. "JivDaya committee." Accessed March 20, 2019. https://www.jaina.org/page/JivDaya

King, M. 1999. *Mahatma Gandhi and Martin Luther King Jr: The Power of Nonviolent Action (Cultures of Peace).* Paris: UNESCO.

Koneswaran, G., and D. Nierenberg. 2008. "Global farm animal production and global warming: Impacting and mitigating climate change." *Environmental Health Perspectives, 116*(5), 578–582.

Lakshmi, R. 2009. "Son marks Martin Luther King's 1959 visit to India." *Washington Post,* February 18, 2009.

Lal, V. 2000. "Too deep for deep ecology: Gandhi and the ecological vision of life." In *Hinduism and Ecology: The Intersection of Earth, Sky, and Water,* edited by C.K. Chapple and M.E. Tucker, 183–212. Cambridge, MA: Harvard University Press.

Larson, G. 2004. "' A beautiful sunset . . . mistaken for dawn', Some reflections on religious studies, India studies, and the modern university." *Journal of American Academy of Religion, 72*(4), 1003–1020.

Lodrick, D.O. 1981. *Sacred Cows, Sacred Places: Origins and Survivals of Animal Homes in India*. Berkeley: University of California Press.

Marriott, M., ed. 1990. *India through Hindu Categories*. New Delhi: Sage.

Mora, C., D.P. Tittensor, S. Adl, A.G.B. Simpson, and B. Worm. 2011. "How many species are there on Earth and in the ocean?" *PLOS Biology* 9(8): e1001127. https://doi.org/10.1371/journal.pbio.1001127.

Nojeim, M.J. 2004. *Gandhi and King: The Power of Nonviolent Resistance*. Westport, CT: Praeger.

Pavapuri. n.d. "Pavapuri—A holy sanctuary for pilgrims." Accessed March 20, 2019. http://www.pavapuri.com/#animals.

Rotman, M. 2010. "Cuyahoga River Fire." *Cleveland Historical*. Accessed April 6, 2019. https://clevelandhistorical.org/items/show/63.

Shree, V. 2018. "Should we drink milk or not?" Accessed May 17, 2019. https://www.youtube.com/watch?v=DLyitfLJ6vk

Vallely, A. 2002. "From liberation to ecology: Ethical discourses among orthodox and diaspora Jains." In *Jainism and Ecology: Nonviolence in the Web of Life*, edited by C.K. Chapple, 193–216. Cambridge, MA: Harvard University Press.

Chapter 6

Respect for Nature

Learning from Indigenous Values

CHRISTINE J. CUOMO

In academic conversations, environmental ethics are often described in terms of "new" values and insights, which gives the impression that the embrace of an ethos of restoration and sustainability involves the discovery or invention of new ideals, rather the revival and cultivation of existing values that have been sidelined and suppressed. The idea that "new" values are what is needed to mend relationships with nature and other species is present in Aldo Leopold's famous characterization of mid-twentieth-century ecological consciousness as the product of the historical evolution of ethics (1949), and it appears in the title of Richard Routley's genre-defining essay, "Is There a Need for a New, an Environmental Ethic?" (1973), in which he argued that indeed there is such a need, due to the complete absence of acknowledgment of the moral value of nature in the history of Western philosophy. More recently, on the topic of virtue, Rosalind Hursthouse has described environmental virtues as new, unfamiliar values that must be discovered and introduced, rather than as traditions to be upheld and reinvigorated (2007, p. 160).

Christine J. Cuomo, *Respect for Nature* In: *The Virtues of Sustainability*. Edited by: Jason Kawall, Oxford University Press (2021). © Oxford University Press. DOI: 10.1093/oso/9780190919818.003.0006

While it is true that environmental ethics are lacking in business as usual under capitalism, and most of the history of (Western) philosophy has either ignored or disparaged values such as respect for nature, portraying environmental virtues as new is misleading, for it denies the existence and relevance of a great range of established ethical systems that recognize the moral value of nature, and depicts respect for nature as a recent development, rather than a complicated and well-established human virtue. Instead of denying or sidelining the presence of biocentric traditions across the globe, philosophers can take seriously the words of the late Native American activist Russell Means, and the values he refers to when he writes,

> I think we should be very happy and proud that we are still Indian people. We're still alive, and we're still resisting. We still have respect for the earth. We have traditional knowledge and values that are superior to anything in Western, "scientific," industrialized culture. Yet industrial society neither understands nor wants our knowledge and our values. That is sad, and we should feel pity for it. (Means, 1996, p. xi).

Russell Means's assessment, published a quarter century ago in the anthology *Defending Mother Earth: Native American Perspectives on Environmental Justice,* edited by Jace Weaver, identifies respect for the earth as a superior value which is alive in indigenous ethics, and which has grounded Native American survival and resistance in the face of the atrocities of colonialism, displacement, and oppression for generations. An aspect of traditional knowledge, respect for earth is considered a virtue and cultural ideal which precedes, exceeds, and competes with hegemonic Western culture and industrial society. It is therefore part of an epistemic and ethical system of interest to those who are dissatisfied with industrial society's values, or lack thereof.

In "Where Do We Go From Here?," his contribution to *Defending Mother Earth,* Thom White Wolf Fassett considers the complexities that emerge for indigenous leaders working in coalition on matters of environmental justice in contexts such as the United Nations. Because indigenous communities face common or similar struggles and adversaries, political strength can be generated together in international organizing. However, Fassett suggests that the real power of such coalitions is found in the collective moral wisdom native leaders bring to the process. Asking "Dare we hope that this new cooperation between Native and the United Nations might begin to provide a basis to defend Mother Earth?" he concludes that hope is found in practices that *restore* values and virtues, and cultivate agreement in light of the common good. Hence Fassett highlights the importance of new cooperations being informed by common or overlapping values that are not so new:

> Only by rediscovering the simplest and most basic values and virtues of the spirit and engaging in creative discourse on the meaning of hope and the nature of the common good will we be able to provide moral precepts and agreements of conduct necessary to create restorative measures for the continued viability of Mother Earth. (Fassett, 1996, p. 190)

The process Fassett recommends is a model for ethical development: thoughtfully and collectively discussing and reflecting on values, virtues, and the common good, to help arrive at and enact agreements that can heal and strengthen social and ecological communities. Virtues are commitments that can provide the grounding needed for this practice, because reflecting on similar or overlapping ethical ideals helps those with differing perspectives develop agreements and common principles. Restored values such as respect for nature can therefore serve collective investment in the continued viability of nature.

MORE THAN A FEELING

The phrase "respect for nature" has long been thought to capture an important and central feature of environmental moral consciousness. In philosophy and the academic field of environmental ethics, this is due in part to the influence of philosopher Paul W. Taylor's theory of "biocentric" or life-centered ethics, presented in his 1986 treatise *Respect for Nature.* Taylor's theory, which builds on the work of the eighteenth-century philosopher Immanuel Kant, is rooted in recognition of the moral value of all living beings. In *Respect for Nature,* Taylor names that recognition "the attitude of respect," which he contrasts with the attitude of exploitation. He describes respect for nature as a "set of dispositions" expressed and embodied in conduct and character: "Ethical action and goodness of character naturally flow from the attitude, and the attitude is made manifest in how one acts and in what sort of person one is" (Taylor, 1986, p. 80). The attitude of respect also leads one to follow rules, such as refraining from harming those with moral value. For example, Taylor argues that, due to the attitude of respect for nature, one might faithfully consider another animal's standpoint and "without a trace of anthropomorphism, (make) a factually informed and objective judgment regarding what is desirable or undesirable from that standpoint" (Taylor, 1986, p. 67).

There are a number of questions to be raised about Taylor's reliance on Kantian ethics, but I will not address those here. Additionally, the radical egalitarianism of Taylor's view of moral considerability is a point of concern. While talk of the basic moral value of all beings and the sacredness of life on Earth may capture important abstract agreements among different ethical communities, the idea that respect for nature is an all-or-nothing endeavor makes it difficult to find common ground with those whose interpretation of that virtue differs from one's own. In addition, a moral view that does not distinguish between different qualities and levels of respect for different objects of respect cannot provide much help as we negotiate real situations rife with legitimate conflicts of interests in relation to natural resources.

Other questions have been raised about Taylor's theory of respect for nature. Is he correct that respect for nature is primarily an attitude? Rosalind Hursthouse has argued that reducing respect for nature to an attitude is mistaken because a virtue must be a deep, substantial character trait. Ideally, a virtue has the ability to be effective in the world, and therefore it must include more than inclination. She therefore characterizes virtues as positive active commitments and informed, motivating traits. They are "efficacious tendencies"—they have active movement and therefore are not mere dispositions (Hursthouse, 2007, p. 163). Here I agree with Hursthouse. Attitudes, which are a blend of feelings, dispositions, and judgments, seem to be a small part of what we refer to when we think of the proactive efforts needed to actualize respect for nature, or when we describe what respect for nature compels when it is a fundamental feature of one's moral perspective. What does it mean to respect a managed forest or a threatened species? At the moment, respectfulness toward threatened species, beleaguered landscapes, polluted bodies of water, and Earth's atmospheric system requires robust protective and restorative actions, policies, and institutional structures. Regarding what is needed for environmental protection, attitudes may even sometimes be beside the point.

Taylor's attitudinal interest may be helpful in describing how respect can empower restraint, or moral actions in which the best action is inaction. However, respect can also require much more than refraining from harming those we respect. And unfortunately, an attitude of respect can be present in situations where it fails to motivate or actualize any positive action. If respect for nature is a virtue, it must consist of more than judgements of worthiness, and it must be more than a feeling.

LAND, AUTONOMY, AND ACTIVE RESPECT

Articulations of indigenous virtues of respect for nature have been developed in depth in a number of works in philosophy focused on ethics, values, epistemology, politics, and metaphysics (e.g., Atleo, 2011; DeLoria, 1972; Kimmerer, 2013; Topkok, 2015; Weaver, 1996; Whyte, 2019). In this body of literature, indigenous philosophers and theorists describe respect for nature as an approach to life that is capable of restoring health, balance, and integrity. However, being respectful also requires tangible effort— sometimes extraordinary effort— and real relationships with animals, plants, land, waterways, and natural elements (Whyte, 2013). Respect is more than an attitude, although the attitudinal aspects of respect are vital. In addition to being felt, respect is enacted in practices and traditions that have been developed over generations, in order to bring a range of beneficial results. Participation in such practices is considered a fundamental right, as well as a basic human function, "just as every other creature has a particular role to play in maintaining the balance of creation as an ongoing process" (Tinker, 1996, p. 171). Personhood itself is therefore built on a fundamental sense of responsibility within mutually supportive social and ecological relationships, and within landscapes where other creatures and living systems are valued for their agency and inherent worth, as well as for their use value as resources for other life forms.

If the virtue of respect for nature includes participating in established protocols in order to maintain good relationships with other beings, it requires the space and freedom to maintain and nurture those relationships. Hence upholding native sovereignty and treaty rights is a crucial form of support for the ethical autonomy of indigenous communities, as well as a matter of political justice. Highlighting the connections between sovereignty and community health, Russell Means writes:

> Our right is to have gardens again! Our right is to have a clean environment on land that we control, where we can hunt, and fish, and gather foods and medicines without fearing that they are too polluted with toxins for us to eat or to use. (Means, 1996, p. xii)

Similarly, historian George Tinker links sustainability to the freedom to live in a culture that is not dominated by colonial modes of consumption:

> What we want is life in the sense of self-sufficient, cultural, spiritual, political and economic sustainability— on our own terms. . . . The question is not whether Indian peoples should have the right to self-determining autonomy, but how our communities can regain this rightful heritage without the continuing colonial pressure to feed the consumptive habits of White America. (Tinker, 1996, p. 170)

Given the recurring, widespread attacks on native peoples' rights to manage and protect their own land and resources, native sovereignty is the crucial element that allows respectful relationships with nature to endure, for tribal sovereignty and land tenure provide the existential, material, and moral context in which respect for nature can be practiced, developed, and realized. As Russell Means writes, "It is our desire to defend our communities and our Mother that makes sovereignty guarantee by treaties the number one demand of indigenous peoples" (1996, p. xi). In line with existing treaties and international law, the guarantee of sovereignty is a straightforward matter of justice for indigenous nations, and for many it is also an avenue for enacting traditional and innovative approaches to continuing, building, and nurturing effectively respectful relationships with/in nature, on their own lands and beyond.

INDIGENOUS VALUES IN CONTEXT: PERSPECTIVES FROM NORTHERN ALASKA

The sense in which the virtue of respect for nature is articulated and engaged in indigenous philosophies and societies is evident in the system of Iñupiaq values espoused by the native peoples of Arctic Alaska. Iñupiaq literature and discursive culture name respect for nature as a core traditional and cultural value, and give life to it as an active practice that inspires actions, shapes priorities, and binds communities together through the cultural practices in which it is expressed, including traditional food preparation, sewing, camping, hunting, berry harvesting, and drumming. Certainly there may be diverse interpretations of the requirements for respecting nature within any community. Nonetheless the Iñupiat have their own unique, complex history of promoting respect for nature as an ethical value within their traditional subsistence culture, but also in the context of participating in intense resource extraction and other environmentally taxing industries.

The ancestral lands and traditional subsistence hunting and camping grounds of the Iñupiat range across the top of the entire state of Alaska, reaching west to the Bering Sea, and east into the immense Arctic Wildlife Refuge that stretches across the Canadian border. Modern Inupiaq communities encompass 34 villages, including the city of Utqiagvik, the economic and political center of the North Slope Borough, located just 9 km from the northernmost point in America. Originally the site of an ancient Iñupiaq settlement, today half the population of Utqiagvik is indigenous (native people comprise 14% of Alaska's total population). In 2016 the residents of Utqiagvik voted to replace the town's former name (Barrow), and restore its original name, which refers to a place to gather wild roots.

Financially dominated by the fossil fuel behemoth since Alaska became a state in 1959, the North Slope Borough has been a significant provider of domestic crude oil and natural gas for decades, with its land and waters long regarded by many interests (public, private, national, international, state, local) as sites for generating profit. On the North Slope, oil, gas, and transport companies, a few of which are locally owned, create a pervasive presence of heavy machinery, dusty roads, and drilling rigs. Those companies are leading employers and philanthropic forces in an remote region hungry for jobs and cash, for the cost of living in northern Alaska is extraordinarily high. Unfortunately they are also contributors to the poor health outcomes that have been well-documented in North Slope communities, including unusually high rates of cancer, depression, addiction, and youth suicide among the Iñupiat (Wernham, 2007; Wexler, 2006).

Even before the wealthy energy and related industries took hold in northern Alaska in the 1970s, the lands and seas of the far north had been subject to extreme environmental intervention and forced change that dramatically impacted local communities, the land, and native species. Before the influence of oil, commercial whaling ships and the phenomenally lucrative whaling industry first brought colonizing impacts to the remote and frozen region of the North Slope, where Iñupiat communities had traditionally and sustainably hunted Bowhead whales for generations. From 1849 through the early 1920s, traditional whaling was supplanted by a massive commercial whaling station operated near Utgiavik. Missionaries established a Presbyterian church in town, and American whaler/settlers brought Eurocentric American architecture and other structures to the North Slope, establishing trading outposts, schools, and government offices, marrying into Iñupiat families, and raising bicultural children (Blackman, 1989).

Although it produced great wealth for the shipping centers of the Lower 48, especially New England and San Francisco, commercial whaling took its toll on traditional native subsistence culture and spirituality, as well as the local food supply, due to its severe impact on bowhead whales and other sea mammals. Yet a century later those species have bounced back due to an international moratorium on commercial whaling and the successful reinvigoration of Iñupiat whaling and sea mammal hunting cultures and practices, in line with the provisions of the Mammal Protection Act of 1972. Nonetheless, troubling questions remain concerning the negative impacts of the fossil fuel economy and its drilling, seismic activities, spills, and pipelines on the people, as well as the sea mammals and other wildlife central to Iñupiaq identities and subsistence practices.

FROM ANCSA TO NPRA-A

In the war years between the end of whaling and the beginning of oil and gas exploration, the US military was also quite active in North Slope, building and operating a huge naval base and an extensive early-warning radar system focused on the threat from the other side of the Bering Strait, and conducting test operations and high-impact exploratory projects over the course of decades, as described in unsettling detail in O'Neill (2015). The establishment of significant military infrastructure on the Arctic coastal plain both foreshadowed and enabled the introduction of the intensive "dirty" resource extraction industries to the remote region. From the Prudhoe Bay oilfields, located at the top of the Trans-Alaska pipeline, to the 23-million-acre National Petroleum Reserve in Alaska (NPRA-A), which surrounds nearly all of the villages of the North Slope Borough, oil and gas development has provided reliable income for the state, the people of Alaska, indigenous communities, and the federal government, which receives taxes from the companies leasing plots of land to search and drill for oil and gas. Rather than "respect for nature," the ethos of oil fields, commercial whaling, and capitalist military installations is one that rewards people for disregarding, exploiting, and damaging nature and other species.

When petroleum and natural gas companies first started working in Arctic Alaska in the 1970s, they did so with the help of the Alaska Native Claims Settlement Act (ANCSA), an agreement which diminished indigenous control of land and resources throughout the state, granting the US government the right to lease and tax land for oil development. In "exchange," Native Alaskan communities were collectively paid a sum of nearly one billion dollars, which was dispersed through the creation of native corporations. With the Settlement Act, the United States therefore secured its right to control the land, and to lease it for exploration and drilling. Agreeing to form corporations provided Native Alaskans with a way to profit from the exploration, drilling, and transport occurring on and near their homelands, and a mechanism for investing in their own long-term economic viability and land security. The one-billion-dollar settlement, signed into law by President Nixon, was paid collectively to Alaska's indigenous people through the creation of big businesses that are owned and controlled by native communities. Hence, as Native Alaskan scholar Jessica Bissett Perea writes, in "a clear departure from the Native American reservation system of the Lower 48, ANCSA stipulated a corporate-to-government relationship between enrolled Alaska Native shareholders and the US federal government," replacing communal land tenure with corporate ownership, and converting village membership into corporate shareholder status (Perea, 2017, p. 146).

The influence of incorporation remains a complicating factor for many indigenous Alaskan communities. Perea finds that "the passage of ANCSA gave the broader Alaska Native community some degree of agency and control over land and resources with which to develop political, economic, and cultural capital," although its negative impacts on identity, health, and economic options for Native Alaskans have been significant (Perea, 2017, p. 147; c.f. Wernham (2007) and Wexler (2007)). She also suggests that by requiring indigenous Alaskans to enroll in corporations for community membership and support, the Settlement Act establishes "a new definition of a 'Native' person that is based on ownership of corporate shares" (p. 146). Indeed, Native corporations have become vocal, if not univocal, representatives of the interests of indigenous Alaskan communities, and forceful protectors of traditional hunting areas, recreational campgrounds, and animal nesting grounds that border or overlap the oil fields. At the same time, incorporated communities' businesses, interests, career trends, and areas of expertise have become deeply invested in and enmeshed with extractive and sometimes seriously harmful industries.

IÑUPIAQ VALUES AND RESPECT FOR NATURE

> The Elders listed our Iñupiaq Values, which we have lived for thousands of years, to officially document what every Iñupiaq should know about our cultural heritage. Listening to our Elders by talking about cultural values helps strengthen the resiliency of our community well-being. (Hogan and Topkok, 2015, p. 67)

Braided into the lived complexities of environmental ethics for diverse and complex Iñupiaq communities, the discourses and practices regarding community morality describe respect for nature as a basic commitment and an active practice. It is a value/virtue that is discussed and endorsed by individuals and collectives, from families and social groups to native corporations. The discourse of Iñupiaq Values (*Iñupiat Ilitqusiat*) is rather unique, as is its historical development.

In the 1980s, across the arctic United States, community elders and leaders launched a "spirit movement" to promote and develop traditional Iñupiaq values, and to strengthen community and cultural health, through renewed commitments to traditional practices including learning and speaking Inupiaq, drumming, dancing, and subsistence hunting. As part of the movement, they also initiated a process centered on codifying the traditional values espoused by elders and community leaders well-versed in the language, stories, and traditions of their people, and which embody the virtues that have enabled the Iñupiaq to thrive and live peacefully together in the demanding context of the Arctic (Napoleon, 1996; Topkok, 2015, p. 12). In a process reminiscent of Thom White Wolf Fassett's call for creative restoration, elders from different Iñupiat villages and regions, including communities based in the North Slope, the Northwest Arctic Borough, Seward Peninsula, and King Island, developed accessible and user-friendly lists of their collective ethical commitments (Topkok, 2015, p. 175). To a visitor to the region today, these values, which are distinct for each community but also overlapping, are visible features of Iñupiaq culture, present in posters, murals, wallet cards, and websites, where they are conveyed in both Iñupiaq and English.

Several scholars have identified a motivation of the spirit movement as the fact that "Iñupiat Elders were concerned that the Iñupiat people, because of the introduction of foreign cultures, were living a destructive lifestyle" (Topkok, 2015, p. 44, citing Napoleon, 1996), including serious concerns about depression, health, and well-being in the community. In addition, there was a robust regenerative aspect of the movement, as Topkok discusses in detail through the lens of his own profound experience as a member of a traditional Iñupiaq dance group. Remembering that the elders and community members who codified these inventories of Iñupiaq values included survivors who had experienced extremely difficult years of cultural assault, language denial, brutality, and trauma from federal and state government, especially in the boarding school system, their accomplishments in restoring Iñupiaq culture are quite remarkable. By reclaiming celebrations and strengthening group structures such as trans-familial whaling teams, as well as traditional sewing arts, cuisine, drumming, dancing, and games, each with its own form of excellence to strive for, the spirit movement of the 1980s was quite effective in restoring the energizing, healthy, and joyful cultural practices that are present in contemporary Iñupiat culture.

In his dissertation "Iñupiat Ilitqusiat: Inner Views of Our Iñupiaq Values," scholar Charles Sean Asiqłuq Topkok, of the Northwest region of Alaska, who describes himself as of Iñupiat, Sámi, Irish, and Norwegian descent, investigates and explicates the Iñupiaq Values of the Kotzebue area of the Northwest region of Arctic Alaska, near the Bering Strait:

> Each Alaska Native group has their own unique set of cultural values, and in some groups the cultural values may differ from village to village. However, all cultural values have been passed down for generations. . . . Elders, community members, and other cultural bearers worked together to identify and create their own cultural values; they did not come from one individual. (Topkok, 2015, pp. 21–22)

Topkok's qualitative research included extensive interviews with 12 indigenous participants as well as an analysis of *Iñupiat Ilitqusiat* as they appear in the prominent Northwest epic story of Qayaq, "an Iñupiaq hero in an unipkaaq (old Iñupiaq oral story)" (Topkok, 2015, p. 19). The story of Qayaq, whose famous journey in search of his brother is traditionally told over the course of many months, is a canonical work in which cultural values are embedded and conveyed (Topkok, 2015, p. 19).

In defining and explaining each of the Kotzebue Iñupiaq values, Topkok draws from the story of Qayaq and analyzes interviewees' personal stories and reflections on Iñupiaq values, how they were learned, and how they contribute to community well-being. The result is an exploration of the meaning of the values and ethical priorities of a culture rooted in cooperative relationships and traditional subsistence practices. In Topkok's words, "the *Iñupiat Ilitqusiat* defines who we are, directs how we should live our lives, and heals us when we are unhealthy" (2015, p. 9).

For Topkok's community, respect for nature is one element in a list of 17 intertwining and interrelated values and virtues: Spirituality, Humility, Avoid Conflict, Knowledge of Language, Hard Work, Humor, Knowledge of Family Tree, Family Roles, Love for Children, Domestic Skills, Hunter Success, Sharing, Respect for Elders, Respect for Others, Cooperation, Respect for Nature, and Responsibility to Tribe. Throughout his work, Topkok emphasizes the interwoven nature of these values, and their underlying ontology:

> Our cultural values constantly converge with one another . . . we see that *Kaŋiqsimauraaliq Irrutchikun* (Spirituality) converges with *Kamaksriliq Nutim Iñiqtanik* (Respect for Nature) when we acknowledge a spirit in all things natural. We see that *Atchiksuaĺiq* (Humility) converges with *Aŋunialguĺiq* (Hunter Success) when we recognize that boastful hunters can be unsuccessful. We see that *Paaqsaaqatautaiĺiq* (Avoid Conflict) converges with

> *Kamakkutiłiq* (Respect for Others) and *Aatchuqtuutiłiq Avatmun* (Sharing) when people share food even though individuals may be feuding. We see that *Ilisimałiq Uqapiałigmik* (Knowledge of Language) converges with *Kamakkutiłiq* (Respect for Others) when we recognize there are different Iñupiaq dialects. We see that *Savvaqtułiq* (Hard Work) converges with *Kamaksriłiq Nutim Iñiqtanik* (Respect for Nature) when we use every part of the animal and tan the hides traditionally. (Topkok, 2015, p. 87)

The values ensconced in the culture and literature of the Iñupiat are a set of exemplary traits and priorities that are interdependent and mutually supportive. According to Topkok, Iñupiat values impart personal and communal virtues that are thought of as ideals and as accessible ways of living. They reinforce communities, traditional subsistence cultures, interdependence, and personal well-being, providing a framework for *Iñuk* or personhood, as well as practical guidance and cultural training. In the ideals encoded in these lists of values, respect for nature is regarded as a positive and beneficial activity that is inextricably tied to other fundamental values and social virtues, such as being a good hunter, sharing, and respecting elders.

Topkok emphasizes that each community's version and interpretation of official ancestral values is unique and locally and linguistically embedded in its context, but also notes similarities in values among indigenous groups worldwide, including humor, respect for elders, respect for nature, spirituality, and sharing (Topkok, 2015, p. 39). He finds especially interesting overlap between Iñupiaq values and the values that have been articulated by Athabaskan peoples (2015, p. 22) as well as those considered central to the Maori and Sami (2015, p. 37).

Topkok also distinguishes among three types of Inupiaq values—those which refer to the person (*Inuk*), those that refer to the responsibilities within families (*Ilat*), and those that are village (*Nunaaqqiq*) values. He categorizes Respect for Nature as a village value, or a value collectively embraced and practiced by individuals, families, and communities. Three aspects or forms of respect for nature are described as central, and these three aspects— respectful learning (from nature), respectful communication (with nature), and respectful harvesting (of nature)— rely on practices which manifest and reproduce respectful attitudes. Here are illustrations from featured interview excerpts:

1. *Learning from nature is shown in respecting nature's wisdom.*

   Oscar and other Elders; they say that what we know, we learn from Nature. Through observations we know what berries to eat, because we've seen the animals eat the right berries. We know which ones are poisonous, because we see the animals not eat those berries. I think we still have to learn from Nature, we still have more to learn from Nature, because the animals have been around a lot longer than we have. I'm sure there's more to learn from them, if only we know how to listen to them, watch them, and observe them. (Topkok, 2015, p. 146)

2. *Communicating with nature is a form of showing respect toward nature.*

   As we become more urbanized, if that's the right word for it, we really lost a lot of our ability to communicate with Nature. And I really think that our ancestors literally were able to communicate with Nature different than we can in our modern world.

   I remember one Elder saying that there're the long ago stories where animals and humans would talk with each other, they're all on an equal plane. That could still happen, but we'd have to speak our Native language to the animal for them to be able to understand us and for us to understand them. At least that's what I've been told. (Topkok, 2015, p. 146)

3. *Respectful harvesting, or not taking more than you need, requires refraining from greed and waste.*

> Well, that's where we get the statement that the caribou gave itself to you . . . No way did you deserve as a human in God's creation. What puts you in a position to kill, that you're also in a position to receive. . . . And if you've respected the caribou and respected the gift of a caribou, and that's the way. And not many white people can even understand you when say, "You went moose hunting?" "Yeah." "Did you get one?" "No, but one gave itself to me, not by any great feat of my own." (Topkok, 2015, p. 145)

Summing up a crucial element of the ethics of harvesting, Topkok emphasizes that "The animal spirits . . . give themselves knowing that every part of their physical body will be used" (2015, p. 165). Understanding how such knowledge and trust are communicated between hunters and animals is moral wisdom imparted by elders, respected hunters, and by the animals themselves.

The themes of practicing virtue through learning from animals, communicating with animals, which includes listening to them, and refraining from greed and waste, are abundant throughout the Qayaq epic—Qayaq even learns about life as a caribou by becoming a caribou (Topkok, 2015, p. 65). Another interesting illustration of Qayaq's relationship with nature is that he receives assistance from animals, even he does not request it, for on their own they consider Qayaq to be a member of their community (Topkok, 2015, p. 25).

COMPLEX POLITICS OF RESPECT

Respect for local animals, and the importance of taking seriously animals' place in the circle of relations, are central in another recent indigenous Alaskan articulation of respect for nature. In 2019 a resolution declaring a "climate change state of emergency" was passed by the Alaska Federation of Natives (AFN), a coalition composed of many native Alaskan communities and organizations (191 federally recognized tribes, 183 corporations, and a dozen consortiums representing native Alaskan communities). Relying on a tradition-centered, contemporary, and future-oriented framing of the importance of confronting the realities of climate change, the declaration includes a specific appeal from indigenous youth:

> We the indigenous youth and future leaders of Alaska are concerned for the survival of our future generations, ways of life, traditional lands, intact ecosystems, emotional, spiritual, and mental well-being due to Climate Change . . .
>
> Many communities across the state face hardships directly correlated with Climate Change, such as the extreme warming temperatures which melt the permafrost, causing mass erosion, resulting in the relocation of entire communities along with devastating the natural habitats of our animal and plant relatives. These impacts have disrupted indigenous seasonal hunting and gathering traditions . . .
>
> In recent years we have lost community members due to unpredictable and unsafe icy conditions, have seen the die off and disease of seals, salmon, migratory birds, shellfish, whales, polar bears, and recognize that these are also our relatives . . .
>
> We, the Alaska Native youth, are asking our tribal leaders to consider, as is traditional, the future of their grandchildren and the generations to come . . . and to declare a state of emergency on Climate Change. (Alaska Federation of Natives, 2019, pp. 123–124)

Alaska Public Media reported that the declaration received significant support from native elders, and the support of elders is also evident in the fact that the resolution emerged from a gathering of native Alaskan youth and elders which preceded the meeting of the statewide Federation (Hughes, 2019). Chief Victor Joseph, chairman of Tanana Chiefs Conference, encouraged the body of the AFN to vote in favor of the measure, with praise for Quannah Chasing Horse Potts and Nanieezh Peter, who presented the resolution to the gathering, "Let's honor these young people who stood up. They need it. And they did it eloquently and in a good way," acknowledging their courage and respectfulness (Hughes, 2019). In expressing and working to institutionalize active caring for their communities, including people as well as the animals they harvest and live among, the work of these young leaders provides another illustration of environmental virtue articulated through practices of respect for nature, and one especially well-suited for our challenging times.

CONCLUSION

Virtues associated with ecological sustainability, such as refraining from overusing natural resources, communicating with nature, and taking responsibility for the environmental impact of our actions, stand in sharp contrast to the practices and policies of the industries, corporations, and power structures that determine many of the details of our lives. Because so many aspects of modern life are driven by forces beyond our influence, it may seem that the desire to actualize environmental ethics is easily reduced to an abstraction or an aspiration. But attitudes, philosophical commitments, and values are not enough to prevent us from contributing to the serious harms presently waged on planet Earth every day. If virtues are to be meaningful, they must entail more than abstract values and internal character traits. Virtues of sustainability must be active, tangible, and effective.

For those raised in societies that discourage respect for nature and accept exploitation as the norm, it may seem easier to protest against exploitation than to embody respectful and regenerative relationships with nature. Structural and cultural forces keep many of us dependent on practices and products that keep us polluting every day, dumping chemicals down the sink, eating meals laced with plastic, and heating our homes with fossil fuel. Even so, the seeds of fully realized respect for nature are present in our protests, our growing ecological consciousness, and our attentiveness to other animals and life in its many forms. This can be so even when we are overwhelmed by the pervasiveness and inescapability of the exploitative cultures that drive our lives. Rejecting exploitation and disapproving the careless degradation of nature emerges from feelings of affinity, empathy, and appreciation, and desires to protect and act respectfully toward the more-than-human world. Virtue compels us to build on that affinity, as well as to protect what we respect.

Rather than characterizing respect for nature as something new, in linking virtue with the idea of *restoration*, Thom White Wolf Fassett characterizes environmental ethics as a return to basic values. Yet he endorses a reflexive, thoughtful rediscovery of wisdom and insight that is contemporary and practical rather than anxiously attached to a romanticized past. He and many others believe such engaged approaches to the rediscovery of basic virtues can enable the development of contemporary moral precepts and agreements of conduct which can nurture ecological and social restoration, and provide for the continued viability of Earth's ecosystems, species, and people. These scholars are writing for their own communities, but also for a multicultural coalition of people who care for Mother

Earth, and each other. Their understanding of respect for nature as a lively, interactive relationship with other beings, land, water, and sky, and a rejection of greed, overuse, and exploitation, provides wisdom and leadership for negotiating a planetary emergency.[1]

REFERENCES

Alaska Federation of Natives. 2019. "2019 Annual convention resolutions 19-56: Declaration for climate change state of emergency in Alaska." www.nativefederation.org/2019/10/2019-afn-resolutions/.

Atleo, E.R. 2011. *Principles of Tsawalk: An Indigenous Approach to Global Crisis.* Vancouver: University of British Columbia Press.

Blackman, M.B. 1989. *Sadie Brower Neakok: An Iñupiaq Woman.* Seattle: University of Washington Press.

DeLoria, V. 1972. *God Is Red: A Native View of Religion.* New York: Grosset & Dunlap.

Fassett, T.W.W. 1996. "Afterword: Where do we go from here?" In *Defending Mother Earth: Native American Perspectives on Environmental Justice,* edited by J. Weaver, 177–191. Maryknoll, NY: Orbis Books.

Hogan, M.P., and Topkok, S. 2015. "Teaching indigenous methodology and an Iñupiaq example." *Decolonization: Indigeneity, Education & Society,* 4(2), 50–75.

Hughes, Z. 2019. "Pushed by young people, AFN declares a climate emergency." *Alaska Public Media,* October 21, 2019, https://www.ktoo.org/2019/10/21/pushed-by-young-people-afn-declares-a-climate-emergency/.

Hursthouse, R. 2007. "Environmental virtue ethics." In *Working Virtue. Virtue Ethics and Contemporary Moral Problems,* edited by R.L. Walker and P. J. Ivanhoe, 155–171. Oxford: Oxford University Press.

Kimmerer, R.W. 2013. *Braiding Sweetgrass: Indigenous Wisdom, Scientific Knowledge, and the Teachings of Plants.* Minneapolis: Milkweed Editions.

Leopold, A. 1949. *A Sand County Almanac: And Sketches Here and There.* New York: Oxford University Press.

Means, R. 1996. "Foreword." In *Defending Mother Earth: Native American Perspectives on Environmental Justice,* edited by J. Weaver, xi–xiii. Maryknoll, NY: Orbis Books.

Napoleon, H. 1996. *Yuuyaraq: The Way of the Human Being.* Fairbanks: Alaska Native Knowledge Network.

1. Thank you to Jason Kawall for your wisdom and patience, and to collaborators, friends, interpreters, and teachers in Alaska.

O'Neill, D. 2015. *The Firecracker Boys: H-Bombs, Inupiat Eskimos, and the Roots of the Environmental Movement*. New York: Basic Books.

Perea, J.B. 2017. "Audiovisualizing Iñupiaq men and masculinities on the ice." In *Critically Sovereign: Indigenous Gender, Sexuality, and Feminist Studies*, edited by J. Barker, 127–168. Durham, NC: Duke University Press.

Routley, R. 1973. "Is there a need for a new, an environmental ethic?" In *Proceedings of the XVth World Congress of Philosophy*, 205–210. Varna, Bulgaria: Sofia Press.

Taylor, P. 1986. *Respect for Nature*. Princeton, NJ: Princeton University Press.

Tinker, G.E. 1996. "An American Indian theological response to ecojustice." In *Defending Mother Earth: Native American Perspectives on Environmental Justice*, edited by J. Weaver, 153–176. Maryknoll, NY: Orbis Books.

Topkok, S. 2015. *Iñupiat Ilitqusiat: Inner Views of Our Iñupiaq Values*. PhD Dissertation, University of Alaska at Fairbanks.

Weaver, J., ed. 1996. *Defending Mother Earth: Native American Perspectives on Environmental Justice*. Maryknoll, NY: Orbis Books.

Wernham, A. 2007. "Iñupiat health and proposed Alaskan oil development: Results of the first integrated health impact assessment/environmental impact statement for proposed oil development on Alaska's North Slope." *EcoHealth*, 4(4), 500–513.

Wexler, L.M. 2007. "Iñupiat youth suicide and culture loss: Changing community conversations for prevention." *Social Science & Medicine*, *63*(11), 2938–2948.

Whyte, K.P. 2013. "Justice forward: Tribes, climate adaptation and responsibility." *Climactic Change*, *120*, 517–530.

Whyte, K.P. 2019. "Way beyond the lifeboat: An indigenous allegory of climate justice." In *Climate Futures: Reimagining Global Climate Justice*, edited by D. Munshi, K. Bhavnani, J. Foran, and P. Kurian, 11–20. London: Zed Books.

PART III

VIRTUES OF SUSTAINABILITY

Chapter 7

Cooperativeness as a Virtue of Sustainability

LAURA M. HARTMAN

> Cooperativeness would be another important characteristic of agents who could successfully address the problem of climate change. Surprisingly, this characteristic appears to be neglected by both ancient and modern writers on the virtues. Perhaps a virtue of cooperativeness is a candidate for creation; or perhaps, though not itself a virtue, cooperativeness would be expressed by those who have a particular constellation of virtues.
>
> —Dale Jamieson, *Reason in a Dark Time*[1]

Philosopher Dale Jamieson says very little else about cooperativeness in his book about climate change. Other environmental philosophers have also mentioned cooperativeness in passing as a virtue with potential.[2] In this chapter, I accept Jamieson's invitation to investigate cooperativeness as a potential virtue of sustainability. I argue that while cooperation itself may be used for good or ill, the virtue of

1. Jamieson (2014), p. 185.
2. See, for example, Kawall (2018), p. 662; Sandler (2007), pp. 42–43; Cuomo (Chapter 6, this volume); and Ferkany (Chapter 3, this volume).

Laura M. Hartman, *Cooperativeness as a Virtue of Sustainability* In: *The Virtues of Sustainability*.
Edited by: Jason Kawall, Oxford University Press (2021). © Oxford University Press.
DOI: 10.1093/oso/9780190919818.003.0007

cooperativeness blends cooperation with other positive qualities of character that amplify it into a public, social virtue. Sustainability of all kinds, particularly social and environmental sustainability, would be enhanced by individuals who exhibit the virtue of cooperativeness. This virtue may be cultivated in practices and structures such as ecological restoration and cooperative business.

I begin the chapter by differentiating cooperation from cooperativeness. Cooperation in itself is not a virtue; it is an action that is common in nature and may have a positive or negative effect depending on the circumstances. I define cooperativeness as a virtue that builds on cooperation to generate positive relationships that enhance public good. Once it is defined and refined, I use specific cases to demonstrate cooperativeness in action: ecological restoration (an example of environmental sustainability) and cooperative business (an example of social sustainability). Throughout the chapter, I draw heavily on resources from Christian ethics, to root cooperativeness's moral standards in a specific, historical tradition. However, multiple religious and cultural perspectives support cooperativeness in one form or another; in that spirit, I hope that this chapter proves interesting and informative to Christians and non-Christians alike.

WHAT IS COOPERATION? WHAT IS COOPERATIVENESS?

The *Oxford English Dictionary* defines *cooperation* as "The action of co-operating, i.e. of working together towards the same end, purpose, or effect; joint operation"; cooperativeness, then, is "the state or quality of being co-operative."[3] Both terms, then, rest on working

3. *Oxford English Dictionary*, 2nd ed., s.v. "cooperation" and "cooperativeness."

(operating) together, but cooperation is an action, whereas cooperativeness is a quality. The earliest uses of these words in the English language relate to religion (*viz.*, whether the Holy Spirit or angels are cooperating with someone), and indeed the root concept is prevalent throughout the Christian New Testament. Paul frequently refers to his disciples and emissaries with the Greek word *synergos* (cooperators). In the gospel of Mark the author uses the word *synergein* (cooperate) to describe God's help to the disciples: "And they went out and proclaimed the good news everywhere, while the Lord worked with them and confirmed the signs that accompanied it" (Mark 16:20, NRSV).[4] Interestingly, one may work together toward a common effect with an entity that is very different from oneself, such as God or nature. In fact, entities that are not human may be said to cooperate (as we shall see later) if their operation brings about a shared end. One may also work together toward a common effect without intending to do so, as in the case of complicity with evil discussed later. Clearly, cooperation itself is too far-ranging and impersonal to qualify as a virtue.

Cooperativeness, however, as a quality of character, could be an interpersonal virtue in the Aristotelian style: one may easily imagine someone being too cooperative with their peers, and therefore acquiescent to the point of self-erasure, or not cooperative enough, and therefore too self-focused and unable to find common cause with others. The virtue of cooperativeness would lie in the context-dependent, appropriate mean between these extremes. It also, however, has something in common with a virtue like justice, in that it requires sound moral judgment: if one cooperates at an appropriate level, but with an actor who has evil or harmful plans, then this

4. For Pauline references to "synergos," see Romans 16:3, 16:9, 16:21; and Philippians 2:25 and 4:3. I am indebted to John S. Bugbee for these insights and references.

quality of character resembles a vice more than a virtue. It is a virtue if it is used for morally positive ends.

Before delving into cooperativeness as a virtue, I first explore cooperation as a concept that is grounded in the interconnections present among humans and in nature. I then explore the other constituent parts of the virtue of cooperativeness: good relationship characterized by respect, patience, and forgiveness; and a generous willingness to put the good of the other, or of the group, before self-centered goals.

COOPERATION: INTERCONNECTIONS AND ETHICS

From an evolutionary perspective, traits like cooperation and altruism may seem anomalous, because scientists have viewed evolution and natural selection as primarily competitive enterprises. However, in recent years, scientists have come to recognize cooperation, altruism, and mutuality as central to the survival of life. Indeed, the earliest forms of single-celled life resulted from the cooperation (literally, working together) of DNA, RNA, and other complex molecules that teamed up in a symbiotic relationship (Early, 2017, p. 192). "Every living thing that aspires to exist at all must master the arts of survival," according to biology writer Colin Tudge, "but the best survival tactic by far, and indeed the sine qua non, is to cooperate" (2013, p. 94). In a similar vein, evolutionary biologist Joan Roughgarden (2009) questions science's emphasis on sexual selection (characterized by competition) over social selection (which requires cooperation). She argues that the latter is equally important. Primatologist Frans de Waal (1997) contends that primates (including humans) naturally seek social connection and peacemaking—we are "good natured." Anabaptist philosopher Christian E. Early writes, "Mother Nature is an experimental pragmatist, and what actually works in the real world, what got life going and keeps it going, is the creativity of adaptive cooperation" (2017, p. 199). Cooperation persists in nature, not necessarily as a deliberate choice, but because it works to keep individuals and groups alive.

The cooperation that exists in nature stems in part from the undeniable interconnectedness of all things, an interconnection that Wendell Berry calls "the Great Economy." This economy "is not the 'sum of its parts' but a membership of parts inextricably joined to each other, indebted to each other, receiving significance and worth from each other and from the whole" (Berry, 1987, p. 72). Thus, Berry sees the holistic dimensions of the earth's economy as indicative of the need to cooperate more than compete. Lutheran ethicist Larry Rasmussen agrees with the intuition of life's interconnectedness; this understanding undergirds a view of the world that requires care and attention to human effects on the nonhuman world. Rasmussen writes that "all the 'createds' [viz., all humans and the rest of creation] are 'relateds' " (1997, p. 29). Those who are related should care for one another and seek mutual benefit, not competition. In Steve Vanderheiden's words, "the fact of our embeddedness in the environment" should lead to "conscientiousness" (Chapter 9, this volume).

That embeddedness manifests as dependence, a topic explored in Alasdair MacIntyre's *Dependent Rational Animals,* in which he reflects on the importance of "virtues of acknowledged dependence" (MacIntyre, 1999). Moral philosopher Mike Hannis has taken MacIntyre's idea of the virtues of acknowledged dependence and expanded this to include "virtues of acknowledged ecological dependence." Hannis writes, "The complex fabric of our interdependent relationships, which as MacIntyre shows is so crucial to the constitution of identity and the development of autonomous character, is woven not only from threads linking human individuals, but also, critically, from those linking each of us to the wider non-human world" (2015, p. 102). These relationships—with other humans and with the rest of creation—constitute who we are. Recognizing this predisposes us to the type of humility, connectedness, and respect that the natural world requires of us.

The Blue River Declaration, written by a collective of philosophers, religious thinkers, writers, and artists in 2012, agrees with this view of dependence and interdependence. They write:

> Given that life on Earth is interconnected, we are called to affirm that all flourishing is mutual and that damage to the part entails damage to the whole. Accordingly, our virtues are cooperation, respect, prudence, foresight, and justice. (Blue River Quorum, 2012, p. 147)

When the Blue River Quorum name cooperation as a virtue that may enhance mutual flourishing, they seem to have in mind a humble attitude toward the human place in an interconnected world. Rather than competing with natural entities, humans can be co-creators (2012, p. 148). This model of human action is neither fully dominant nor unrealistically submissive but rather helpfully mutual and based in partnership. Deliberate cooperation with life-giving, positive forces may exemplify one form of good human agency in an interconnected world.

Cooperation characterizes the natural world, and if humans are to play a harmonious role in that world we, too, should cooperate with nature. But while nonhumans engaging in symbiosis or mutualism seems innocuous enough, human cooperation—particularly human cooperation with other humans—is not always morally good. When it comes to the human realm, I seek to valorize *willing*, not *coerced*, cooperation; and *mutually respectful*, not *oppressive* cooperation. At its best, cooperation might create harmony between individual and group needs: it provides a model for an individual to thrive in, and contribute to, a collective entity without being swallowed up by it. Additionally, cooperation prompts an individual toward virtue if it entails cooperation *with good*. In Gandhi's words, "Non cooperation with evil is as much a duty as is cooperation with good."[5] It is rarely as simple as he makes it sound, however, to cooperate with good and avoid cooperation with evil.

Catholic moral theologian Julie Hanlon Rubio explores the nuances of cooperation with evil, a technical concept from Catholic moral theology that entails "help afforded another . . . to carry out his purpose of sinning" (2011, p. 104). Rubio focuses on cooperation with social sin, including environmental degradation, sweatshop labor, and other oppressive relationships characteristic of a dehumanizing, exploitative economy (2011, p. 110).[6] Social and environmental sustainability does not support social sins like this. But how responsible is each person who "cooperates" with these practices? On an individual level, living a "regular" North American life that includes driving a car, wearing typical clothing, and shopping at a standard grocery store involves complicity in these types of social sins. Rubio concedes, "We cannot avoid all connection with evil, and those who seek total nonparticipation face certain frustration as well as limited opportunities to do good amid imperfection" (2011, p. 107). Indeed, some may say "it is sometimes necessary to tolerate material cooperation for the sake of a greater good" (Rubio, 2017, p. 112).

Still, after humanity has lived through many atrocities—from colonialism to the Holocaust to more recent genocides—it becomes harder to shrug off everyday participation in "banal" but significant evil (Rubio, 2017, p. 112). Rubio draws on womanist theologians to

5. Quoted in Rubio (2011), p. 103.

6. For her argument, and mine as well, let us bracket the nuances of whether sweatshop labor is 100% evil. The term can be a stand-in for any objectionable practice in a supply chain.

assert the importance of acting "to avoid cooperating with the social structures that harm vulnerable persons whose connections to me are undeniable" (2017, p. 111).[7] For Rubio, this includes avoiding buying clothing made in sweatshops since doing so is substantially, materially cooperating with evil (2011, p. 115).[8] This is not the end of the story, however. The womanist theologians who inspire her suggest "that disentanglement from morally problematic situations may be less important than participation in actions designed to alleviate suffering" (2017, p. 116). In other words, perhaps we should focus less on cooperation with evil and more on cooperation with good—an emphasis that also aligns with this chapter's interest in the virtue of cooperativeness.

To examine cooperation with good in a context of concern about sweatshop labor, Rubio interviews Catholic Workers in St. Louis, who buy used clothing.[9] She observes that "by buying used clothes, they cooperate with good, making an environmentally friendly and often charitable choice" (2011, p. 114). In their own words, they justify their choices as follows:

> The most pure and effective resistance is in the creation and attempted execution of a viable alternative. . . . A lifestyle that, if everyone were to join us, would reverse the destructive spiral of the dominant culture and land us instead on the path toward justice, peace, sustainability, holistic health, and, ultimately, love. (Rubio, 2011, p. 114)

7. Rubio's womanist inspirations include African American theologians M. Shawn Copeland, Emilie Townes, Delores Williams, and Jamie T. Phelps.
8. These are technical terms: material cooperation is contrasted with formal cooperation (it is unintended, not intentional); substantial is contrasted with mediate (it is physically present and linked to the evil, not distant from it).
9. The Catholic Worker is an organization, founded in the 1930s by Dorothy Day and Peter Maurin, that seeks solidarity with the poor through voluntary poverty. Catholic Workers typically run houses of hospitality, soup kitchens, and the like.

By cooperating with good, they are helping to bring about a better world. Philosopher Peter Wenz would commend this impulse to live now a vision of the future, and he has coined the term "Principle of Anticipatory Cooperation (PAC)" to describe this phenomenon. "The PAC," he explains, "calls for actions that deviate from the social norm in the direction of the ideal that virtuous people aspire to for themselves and others but which do not deviate so much that virtue impairs instead of fosters flourishing" (2005, p. 211). Wenz posits that cooperation with good—taking on ecologically friendly habits that anticipate the world we want to live in—constitutes a duty for the ecologically virtuous.

To sum up, cooperation seems to be a characteristic of nature, since all creatures' interconnectedness with each other leads them to at least some cooperation in order to survive. Humans would do well to recognize their dependence on each other and on nature; this state of dependence creates a motivation to cooperate. Among humans, cooperation may also be deliberate or accidental, it may arise from better or worse relationships, and it may result in better or worse outcomes. We see the beginnings of the virtue of cooperativeness in calls to seek action that lies on the "better" side of these spectra.

COOPERATIVENESS: RELATIONAL, PUBLIC VIRTUE

While cooperativeness may simply be "the state or quality of being cooperative," I wish to establish a more robust definition of cooperativeness *as a virtue*. To be a virtue, cooperativeness must exhibit an excellence of character that enhances the virtuous person's ability to live a flourishing, fulfilled life while also enhancing the good of the person's community (since, as we have argued earlier, our deep interconnections lead us to dependence on one another in order to flourish). Individuals who exhibit the virtue of cooperativeness (a) know how to create harmonious relationships and (b) make sure those relationships move all involved parties toward what is good for the group.

Cooperativeness in relationships. For some thinkers, the opposite of cooperativeness is competitiveness; for others it is individuality. In both cases, cooperativeness contrasts with an oppositional self-centeredness. Cooperativeness means recognizing what is good for another and often what is good for a larger group; those who exhibit the virtue of cooperativeness prioritize the community and move beyond their own narrow self-interest. They seek to work together for mutual aid and the common good, concepts that resonate with basic Christian virtues such as generosity and neighbor-love.

However, such a characteristic is not a clear fit for classic virtue theory, which emphasizes personal eudaemonia. In fact, some Christian thinkers have doubts about all of virtue ethics for this reason. Jennifer Herdt writes, "an ethic of virtue is seen as problematic insofar as it fosters a focus on self-cultivation rather than on love of neighbor" (2008, pp. 345–346); pursuit of virtue is only acceptable if the virtues are contextualized as gracious gifts from God and used to help the neighbor. However, cooperativeness might be classified as a burdened virtue, in Lisa Tessman's words—one of the few virtues whose benefits fall on the larger group more than on the virtuous individual (2005).

Cooperative individuals "play well with others," as the common elementary school phrase describes it. They know how to create and maintain good relationships. One factor is patience (explored in greater depth by Kawall, Chapter 8, this volume). Cooperativeness, sometimes framed as "reciprocal altruism," should come more easily to those who can forego "an immediate benefit for the sake of a greater long-term reward" since cooperation means prioritizing someone else's needs over, or at least equal to, one's own (Curry, Price, and

Price, 2008, pp. 778, 781). Evolutionary anthropologist Oliver S. Curry and evolutionary psychologist Michael E. Price found this hypothesis to be true: "patient people are indeed more cooperative." In fact, "some theorists have suggested that the evolution of 'patience' preceded, and was a necessary pre-requisite for, the evolution of reciprocal altruism" (Curry, Price, and Price, 2008, p. 781). In other words, cooperativeness depends on the virtue of patience.

Cooperation also appears to depend on forgiveness in scientific literature. Evolutionary researchers Benjamin M. Zagorsky et al. found that a "forgiver" strategy in the classic prisoner's dilemma[10]—a strategy which mirrors the opponent's cooperation or defection, while still "forgiving" or returning to cooperation even after defection—"facilitates stable cooperation in the presence of exploitation" (2013, p. 1). Forgiveness keeps the relationship cooperative and stable, which appears to favor its evolutionary continuance.[11] Since cooperativeness entails relationship, cooperative individuals must have the virtues that help to create and foster good relationships.

A cooperative individual cooperates with some *other* and this requires other-focused virtues. Awareness—in Cheryl Hall's words, "emotionally registering something (or several things) as valuable"—is an antidote to a self-centered outlook: one must be aware of and

10. The prisoner's dilemma sets up a situation as follows: two individuals are captured and accused of a crime. If neither confesses, they go free. If both confess, they go to prison. If one confesses, the confessor goes free while the other is sent to prison. The prisoners are separated and have to guess what the other is likely to do. The best outcome is for the prisoners to cooperate—for both to not confess and go free. But the temptation to "defect"—to confess while sending the partner to prison—is strong. In this experiment, the researchers set up multiple iterative prisoner's dilemmas that followed different strategies based on the other's actions. Many environmental issues resemble the prisoner's dilemma in that they seem to demand a certain sacrifice on the part of the individual (person or nation) which only truly solves the environmental problem if others also undertake a similar sacrifice. The temptation to keep polluting or using resources is strong because others, it seems, are not going to do their part.

11. For more on forgiveness as a virtue of sustainability, see Hall, Chapter 4, this volume.

open to the other in order to cooperate with them (Chapter 4, this volume). Respect moves beyond awareness to a recognition of the worth and value of the other: a cooperative individual properly values and respects the other. This respect should, of course, be mutual. Christian ethicists Diane Yeager and Stewart Hermann make a similar point about compromise: they write that "the recognition of the worth and moral competency of the other" must be mutual and ideally based on trust (2017, p. 12). Cooperativeness shares with compromise the necessity of awareness and respect of the other, as well as a will to the good for the other.

Cooperativeness wills the good, in a context of mutual aid and help; those who cooperate wish each other well. Indeed, as we shall see later, Japanese evangelist Toyohiko Kagawa saw cooperativeness as a manifestation of Christian love—the highest form of benevolence in Christian thought. Aristotle understood benevolence as a political virtue, a mean between obsequiousness and churlishness; it involves a proper concern for the well-being of another.[12] In sum, cooperativeness rests on good relationships characterized by an appropriately virtuous other-orientation: those who are too self-centered, whose nature is too "curved in upon itself at its deepest levels," cannot effectively cooperate (Luther, 1961, p. 159).

Cooperativeness as public and collective. According to philosopher Brian Treanor, environmental virtues must include, and indeed foreground, "public virtues," which "benefit the community rather than the individual" (2010, p. 10). Specifically, "Public virtues (1) contribute primarily to the well being of the community, though there may (2) be an element of direct personal benefit to having the virtue, and there is always at least some indirect personal benefit insofar as the individual is a member of the community" (2010, p. 13). As examples he cites neighborliness, emplacement, and political engagement.

12. As discussed in Brian Treanor (2010, p. 24).

I would add cooperativeness to the list since a cooperative individual willingly sets aside their own well-being for the good of the whole or the other—this, too, contributes to the community.

Treanor reasons that since environmental problems result from collective action, the solutions must also arise from the collective. Jamieson's treatment of climate change, which inspired this chapter, points to the need for collective change and the difficulty of marshalling such action under current circumstances. But public virtues provide hope, according to Treanor. He writes, "public virtues are character traits that bring us into virtuous relationships with our communities and environments," and in so doing they encourage public engagement that makes a positive difference (2010, p. 18).

Cooperativeness's status as a public virtue connects with its social nature: it is an other-facing virtue and can only be practiced with another. Catholic moral theologian Meghan J. Clark describes social virtues in similar terms: "While other virtues . . . may focus on the habits and agency of the one, social virtue must understand agency in terms of participation" (2014, p. 30). As a social virtue, cooperativeness responds to, and may be embedded within, social structures. Accordingly, moral theologian Daniel J. Daly argues that structures—both the "objectified social structures" as well as "socially rooted moral habits" in individuals—may be judged as virtuous or vicious (2011, p. 355).[13] Following Daly, Clark speculates "how a community or social group can cultivate and habituate the virtue of solidarity" or another virtue such as cooperativeness. She concludes that "doing so requires both the cultivation of practices that habituate solidarity [or cooperativeness] as well as structures of

13. Daly defines a structure as "an institution, a practice, a value laden narrative, or a paradigmatic figure . . . which orientates or organizes economic, social, and political life." Both ecological restoration (a practice) and cooperative businesses (institutions) would count as structures according to this definition.

virtue" (2014, p. 35). What practices or structures might habituate the virtue of cooperativeness, allowing a community or social group to cultivate this virtue? I explore two possibilities next.

PUTTING COOPERATIVENESS INTO PRACTICE

Christian environmental ethicist Willis J. Jenkins writes that ethical responses to climate change and other environmental ills should be "grounded in local response and . . . pragmatic"; he seeks models of action that are at the same time "prophetic" and "pragmatic." This is the way to "create cultural transformation through critical participation in reform processes" (2013, p. 9). The two examples that follow, ecological restoration and cooperative business, both represent prophetic, pragmatic reforms that have important implications for social and environmental sustainability. These are within reach of many everyday people—volunteering to help with an ecological restoration project is not too onerous for most; planting native plants in one's own yard, or advocating for ecological restoration in public parks is also fairly easy to accomplish for many people. Similarly, cooperative businesses are relatively easy to find (credit unions, food co-ops, electric cooperatives) and participate actively in. As I argue later, active participation in these practices and institutions builds the virtue of cooperativeness, which is vital to create a more sustainable future.

Ecological Restoration

The Society for Ecological Restoration defines *ecological restoration* as "the process of assisting the recovery of an ecosystem that has been degraded, damaged, or destroyed" (2019). It involves humans in actions that reverse or repair human-caused damage: planting trees after a clear-cut, daylighting a stream that had been buried in a culvert, or eradicating invasive species so that wildflowers might flourish in their place. A hopeful, often volunteer-based activity, ecological restoration casts humans as healers. But in most cases humans seek to restore something akin to "wild" nature—thus the anthropogenic, intentional dimensions of this activity exist in tension with the value of letting nature take its course. Many restorationists seek to create an ecosystem that appears natural, even though it was assembled by benevolent humans in the wake of harmful human action. Philosopher Eric Higgs recognizes the paradox of human agency in restoration when he writes, "Intentionality is critical to the process of restoration, but intentionality courts hubris" (2003, p. 285). We risk inflating our sense of power: can we really create an ecosystem from its constituent parts? Benevolent though it may be, ecological restoration also implies inflated power for human actors. For this reason, Higgs writes, "I propose that we think of *restoration as conversation*" (p. 285). This implies "reciprocity" between humans and the ecosystem in question, a process orientation characterized by "continuity over time" and "participation" between humans and the natural world; in this way, "the interests of ecosystems and people are both engaged" (p. 286).

Higgs does not explicitly use the word *cooperation*, but Christian ethicist Gretel Van Wieren does. "Restoration," she writes, "is fundamentally about human beings learning to actively participate in and cooperate with nature's inherently wild ways" (2013, p. 42). Van Wieren sees restoration as humble, not hubristic. "Restoration is not an inherently egotistical activity," she asserts; "it is not playing God, it is playing human. To be human, bioregional restorationists believe, is to be active participants, givers and takers, reciprocators, loyal dwellers in and with the land" (p. 46). This echoes the words of conservationist Aldo Leopold, who advocates a "land ethic" premised on human interconnection with and not just cooperation with but cooperativeness toward the land system. Leopold writes, "a land ethic changes the role of *Homo sapiens* from conqueror of the land-community to plain member and citizen of it. It implies respect for his fellow-members, and also respect for the community as such" (1970, p. 240). As one of the originators of ecological restoration in North America (he famously established the first restored prairie in Madison, Wisconsin), Leopold would agree with Van Wieren's assessment. An ethical relation to the land means cooperativeness: a relationship of mutual service, reciprocity, and harmony.

Van Wieren sees the act of restoration as holy. "Where creation's given sanctity is damaged by human activities and can still be put back by human activities, it should be—as an act of obedience, service to, or friendship with the God who created it in the first place" (2013, p. 69). She implies that one is not only cooperating with nature but, perhaps, cooperating with God as well. This echoes those early uses of *synergos* in the Christian New Testament, in which the Holy Spirit cooperates with early evangelists. Van Wieren goes so far as to wonder if restoration may constitute forgiveness for ecological sins: when the land regenerates, it seems to forgive humans for its wounds (pp. 96, 178–179). In the process, humans become "re-centered as partners with and participants among, rather than stewards over, the rest of creation" (p. 105). Cooperating with nature puts us in our place, in both senses of the phrase; it also demonstrates what ought to be, in a kind of public witness to the world (2013, p. 163). Wenz would probably say that ecological restorationists are enacting the PAC, since their actions anticipate, and help to bring about, a more ecologically harmonious world.

Ecological restoration has its detractors, among them philosopher Eric Katz. Restoration is a misnomer, he argues; we restore nothing. At best, humans create a nearly perfect substitute for a destroyed ecosystem, but it should not be mistaken for anything "natural": it is an artifact. "Restoration ecology," he writes, "is a continuation of the paradigm of human scientific and technological mastery over natural processes. The underlying technological assumption is that humans can control natural processes to better effect than nature can" (Katz, 2015, pp. 489–490). Van Wieren would disagree—her version of ecological restoration promotes a humble, cooperative approach. But Katz would consider Van Wieren naïve and idealistic; it is a slippery slope between "cooperation" and "domination," especially when humans (who Katz believes have a tendency to seek control) interact with nature (which we have sought to control for millennia—and largely succeeded). Arguably, that human impulse to domination is the reason we must deliberately cultivate the virtue of cooperativeness: to proactively seek a more mutual, life-giving relationship with nature than the one we currently have.

Cooperatives

Cooperatives are businesses with a distinctive ownership structure: they are owned collectively, by the workers or consumers, for their own mutual benefit, instead of being owned by shareholders for the sake of profit.[14] Cooperative models for business and mutual aid lend themselves to social, economic, and even environmental, sustainability, because of the way the business model works. This model of business organization involves equality among decision makers ("one member = one vote") by contrast with a shareholder

14. Cooperatives originated with the Rochdale Equitable Pioneers' store, a consumer cooperative in Rochdale, England, in 1844. Some of the guiding principles include "voluntary and open membership," "democratic member control," "member economic participation," "cooperation among cooperatives," and "concern for the community." In the United States cooperatives flourished in the Depression era and in the 1960s and 1970s, but even today many common institutions, such as food cooperatives, credit unions, and utility cooperatives, are still organized using these Rochdale principles. (Cooperative Education Network, "Rochdale Pioneers," "Rochdale Principles," and "History of Cooperatives in the United States.")

model (more shares owned = more votes). Typically, profit made by the business is reinvested in the cooperative or shared equitably among members, thus enhancing both social and economic sustainability.

Cooperatives have received widespread support from multiple sectors of society. Pope Francis has praised cooperatives repeatedly, and indeed much of Catholic Social Teaching seems to support both the virtue of cooperation and the formation of cooperatives (Danner, 2016, p. 188n1).[15] According to Christian ethicist Kerry Danner, African Americans have historically found multiple ways to engage in cooperation, whether from informal arrangements or formalized, incorporated cooperative businesses.[16]

Cooperatives promote social and environmental sustainability. As a business focused on equity and supporting the community, cooperatives strengthen community ties and promote resilience. Credit unions, a form of cooperative, typically invest in local business, contributing to economic and social sustainability. The majority of food cooperatives promote environmentally conscious consumption; consumer cooperatives tend to reduce waste by buying in bulk. The Wendell Berry Center in Kentucky facilitates a beef cooperative to help farmers who are raising their cattle in an environmentally friendly manner.[17] Danner writes, "Insofar as certain types of economic cooperatives resist social sin and ecological harm and realign our behavior with a desire to heal and protect creation, they offer an

15. This note in Danner's paper lists the many times Pope Francis has mentioned cooperatives approvingly.

16. The history here is fascinating. Consider one example, Jacob Reddix of Gary, Indiana, who responded to the Great Depression by founding the Consumers' Cooperative Trading Company (1932) and went on to teach classes on cooperation, creating "an ice cream and sweets co-op run entirely by children, two large grocery stores, a credit union, and a gas station—all functioning cooperatives" (Danner, 2016, p. 179).

17. Berry himself has a history with agricultural cooperatives as well. These local-based structures of mutual aid appeal to the agrarian values that he advocates (The Berry Center, n.d.).

ethically defensible mechanism in the face of climate change" (2016, p. 176). Given that consumer capitalism as we know it is on a crash course with various ecosystem limits, including the climate, any alternative—including cooperatives—has potential to defuse looming economic-ecological catastrophes.

Danner also recognizes the role cooperatives can play in cultivating engaged citizenship among their members. Membership has been shown to increase civic engagement more broadly, which "paves the way for a potential increase in wide scale collective action, the likes of which may be increasingly necessary to rein in corporate and industrial power to curb the effects of climate change" (2016, p. 183). It would seem that the act of participating in a cooperative—the act of cooperating—increases civic virtues necessary to combat climate change. Treanor would appreciate this effect of cooperatives. In advocating for the importance of public virtue, he writes, "today, the need for political activism is so urgent, so dire, that it fundamentally changes the kind of person we need to be in order to flourish" (2010, p. 27). True environmentalists, according to Treanor, need virtues of civic engagement—including cooperativeness—in order to meet the challenges of our time.

Cooperative enterprises are not guaranteed to succeed; personal virtues of cooperativeness are required. Individuals seeking to cooperate need what Ferkany and Whyte call "participatory virtues," emphasizing "inclusiveness and engagement" as primary to a productive process. They also argue for "basic self-confidence" (as a mean between arrogance and crippling self-doubt), "reasonableness and fairness," "deliberative friendliness, sincerity, empathy, and charity," and "patience and resilience" (Ferkany and Whyte, 2012, 429–430). This set of virtues echoes the qualities constitutive of cooperativeness. While many cooperative businesses enjoy long lives and happy members, some cautionary tales are in order because of the light they shed on cooperativeness as a virtue.

Reinhold Niebuhr, a highly influential twentieth-century Christian ethicist, participated in a cooperative farm for former sharecroppers during the Depression era. He and some idealistic seminarians bought land in the Mississippi delta in 1936, seeking to turn it into a cooperative enterprise that would help struggling African Americans. In framing the co-op's principles they wrote that cooperation "may be one important non-violent method by which we may pass from an old, outworn order to something new and better" (Byrd and Loucky, 2016, p. 77). This idealism proved short-lived, however. Through a combination of local opposition, inexperience, bad weather, and bad management, the Delta Cooperative folded within a few years. The bad management, according to historical reports, stemmed in large part from white leaders of the cooperative who did not allow black members' voices to be fully heard or their views to be considered (Byrd and Loucky, 2016, p. 78). The cooperative principles include democratic member control, but this was imperfectly executed in the Delta Cooperative. In other words, a cooperative requires members and leaders to exhibit the virtue of cooperativeness—including attention, respect, and benevolence. Without this, the cooperative fails.

Reinhold Niebuhr had his doubts about cooperatives from the beginning. In 1936, as Delta Cooperative was being established, Niebuhr also met Japanese evangelist Toyohiko Kagawa, a well-known leader in Japan's cooperative movement. Kagawa was four times nominated for the Nobel Peace Prize for his work among Japan's poor; he established various cooperatives to aid impoverished workers and struggling farmers. Niebuhr "found Kagawa's hopes for cooperatives unrealistic, based on illusions about human nature and political and economic realities" (Byrd and Loucky, 2016, p. 64). Niebuhr did not hesitate to criticize Kagawa in print.

Kagawa's idealism, however, ran strong. Kagawa felt that the Christian God—who is love and who defines love and who commands love—required working on behalf of others in a cooperative, mutual fashion (Kagawa, 1936a, p. 37). In *Brotherhood Economics,* he advocated the work of "brotherly" love in the form of cooperative organizations, since this entailed mutual benefit and a minimization of harmful competition. For Kagawa, Christianity calls for "conscious acts of redemptive love" that foster cooperation rather than competition (1936a, pp. 195–196). God's abundant love should prompt Christians to set up cooperative businesses, and indeed an entire cooperative society, making it possible to "bridge over the gap between producers and consumers with brotherly love" (p. ix). Kagawa believed that cooperatives could harmonize individual desires and community needs.[18] Although he greatly respected Niebuhr, Kagawa was not dissuaded by his criticisms (Byrd and Loucky, 2016, p. 75).

However, it should be noted that Kagawa's own interest in cooperatives blinded him to evil. He "tragically linked his romantic vision for Christian cooperatives with Japanese militarists' strategies"; he "cooperated" with military authorities by "recruiting and sending 250 Christian 'pioneers' to build 'a new Japan' in Japanese-occupied Manchuria in 1944–1945" (Byrd and Loucky, 2016, p. 83). Japan had invaded and occupied Manchuria years earlier and the area remained a site of many casualties and war crimes; for an acknowledged Christian pacifist such as Kagawa to cooperate with this occupation does appear to qualify as "cooperation with evil." This historical note serves as a useful reminder that Niebuhr's criticisms of Kagawa, which included the charge that too much idealism becomes idolatry, blinding one to the limits of one's preferred vision, may have been correct (Byrd and Loucky, 2016, pp. 74–75).

18. He writes, "The most important problem facing economics is, therefore, how to harmonize these two, freedom and brotherly love. Here enters the cooperative movement" (p. 93).

Nevertheless, Daly's insights about structures of virtue and vice may help to analyze both Niebuhr's failed cooperative and Kagawa's compromised morality. I argue that the cooperative structure is a virtuous structure because it is set up "to promote the human good and human happiness" and to "socially root [positive] moral habits" (Daly, 2011, p. 355). However, in the case of Delta Cooperative, Niebuhr and colleagues were wrestling with an incredibly strong structure of vice—American racism, both in rural Mississippi where the farm was located and in the hearts and minds of the elite Yankees attempting to run the cooperative. Similarly, Kagawa had significant structures of virtue to draw on—both cooperatives and the structures of progressive, charitable Christianity. But when he encountered an overwhelmingly strong structure of vice—the militarism of wartime Japan, bolstered by propaganda—he became blind to the evil with which he cooperated. Pope John Paul II, according to Daly, recognized "that while a person can transcend his social order, he is conditioned by it, and educated within it. Thus, a society marked by unjust institutions and conditions is a society that induces persons to lives of sin, and vice. In such a society heroic moral effort is required in order to avoid the acquisition of vices" (p. 355).

That said, Wenz's PAC comes into play here as well. Even if a society is racist, militaristic, or otherwise blameworthy, virtuous individuals are called to participate in—and in some sense create—a better society. Whether we cooperate with nature's patterns by planting trees to restore denuded soil, or cooperate with one another to create a mutually beneficial business, we are cooperatively bringing about something better. Even if ecological restoration veers toward the controlling more than the cooperative; even if cooperative businesses rest on imperfectly virtuous individuals conditioned by the vices of their time; nevertheless both of these move participants, by greater or lesser amounts, toward the public virtue of cooperativeness.

CONCLUSION

Context and social structures can set up individuals to embrace virtue or avoid vice—or they can do the opposite. Perhaps practices such as ecological restoration and institutions such as cooperatives exemplify this insight. These virtuous structures foster, among others, the virtue of cooperativeness. Cooperativeness, as I have argued, is a complex virtue drawing on multiple other virtues: attentiveness, respect, benevolence; virtues of good relationality such as patience and forgiveness; community-mindedness; and justice. It is both an individual and a social or collective virtue, and it can be fostered by structures of virtue.

Toyohiko Kagawa, for all his faults, was a prolific fiction author and was twice nominated for the Nobel Prize in Literature. One of his best sellers, *A Grain of Wheat,* depicts a program of rural renewal for mountainous Japan in a time of economic struggle, and it is surprisingly ecological in its focus. In the book, the main character, Kakichi, is drawn to Christianity when he listens to someone read the Bible and his mind immediately conjures images of nature's beauty. The land itself might be described as a character in the novel, since the barrenness of the slopes causes poverty and starvation in the main characters. A Christian hermit living in the hills diagnoses the problem: the people do not fully appreciate the land's potential because it does not conform to their image of farmland; if they listened to and appreciated the land, they would learn how it could support them, and they all could be "God's children." The story's happy ending comes from the creation of rural cooperatives, but this only works because the villagers have (cooperatively) replanted the barren slopes with trees, restoring the local ecosystem and making a (cooperative, collective) living from non-timber forest products such as nuts. The story neatly brings together multiple elements: cooperating with the land, cooperation among humans, and cooperating with God (Kagawa, 1936b).

It is a romanticized tale, but instructive nonetheless: cooperation can and should function as a virtue of sustainability.

REFERENCES

Berry, W. 1987. *Home Economics*. New York: North Point Press.

The Berry Center. n.d. "Our home place meat." Accessed June 29, 2019. https://berrycenter.org/projects/home-place-meat/.

Blue River Quorum. 2012. "The Blue River Declaration: An ethic of the earth." *Journal for the Study of Religion, Nature, and Culture, 6*(2), 146–150.

Byrd, B.G., and J.P. Loucky. 2016. "Toyohiko Kagawa and Reinhold Niebuhr: The church and cooperatives." *Journal of Interdisciplinary Studies, 28*(1–2), 63–88.

Clark, M.J. 2014. "Anatomy of a social virtue: Solidarity and corresponding vices." *Political Theology, 15*(1), 26–39.

Cooperative Education Network. n.d. a. "History of cooperatives in the United States." Accessed June 29, 2019. http://cultivate.coop/wiki/History_of_Cooperatives_in_the_United_States.

Cooperative Education Network. n.d. b. "Rochdale pioneers." Accessed June 29, 2019. http://cultivate.coop/wiki/Rochdale_Pioneers.

Cooperative Education Network. n.d. c. "Rochdale principles." Accessed June 29, 2019. http://cultivate.coop/wiki/Rochdale_Principles.

Curry, O.S., M.E. Price, and J.G. Price. 2008. "Patience is a virtue: Cooperative people have lower discount rates." *Personality and Individual Differences, 44*(3), 778–785.

Daly, D.J. 2011. "Structures of virtue and vice." *New Blackfriars, 92*(1039), 341–357.

Danner, K. 2016. "Hope, courage, and resistance during climate change: Insights from African American economic cooperative practices." *Journal of the Society of Christian Ethics, 36*(2), 173–192.

De Waal, F. 1997. *Good Natured: The Origins of Right and Wrong in Humans and Other Animals*. Cambridge, MA: Harvard University Press.

Early, C.E. 2017. "The risky adventure of *Homo Caritas*: The evolutionary story of adaptive cooperation and love." *Conrad Grebel Review, 35*(2), 182–200.

Ferkany, M., and K.P. Whyte. 2012. "The importance of participatory virtues in the future of environmental education." *Journal of Agricultural and Environmental Ethics, 25*, 419–434.

Hannis, M. 2015. "The virtues of acknowledged ecological dependence: Sustainability, autonomy, and human flourishing." *Environmental Values, 24*(2), 145–164.

Herdt, J. 2008. *Putting on Virtue: The Legacy of the Splendid Vices*. Chicago, IL: University of Chicago Press.

Higgs, E. 2003. *Nature by Design: People, Natural Processes, and Ecological Restoration*. Cambridge, MA: MIT Press.

Jamieson, D. 2014. *Reason in a Dark Time: Why the Struggle Against Climate Change Failed and What It Means for Our Future*. New York: Oxford University Press.

Jenkins, W.J. 2013. *The Future of Ethics: Sustainability, Social Justice, and Religious Creativity*. Washington, DC: Georgetown University Press.

Kagawa, T. 1936a. *Brotherhood Economics*. New York: Harper & Brothers.

Kagawa, T. 1936b. *A Grain of Wheat*. New York: Harper & Brothers.

Katz, E. 2015. "Geoengineering, restoration, and the construction of nature: Oobleck and the meaning of solar radiation management." *Environmental Ethics*, 37(4), 485–498.

Kawall, J. 2018. "Environmental virtue ethics." In *The Oxford Handbook of Virtue*, edited by N. Snow, 659–679. New York: Oxford University Press.

Leopold, A.1970. *A Sand County Almanac*. New York: Ballantine Books.

Luther, M. 1961. *Luther: Lectures on Romans*. Trans. and ed. by Wilhelm Pauck. Phil adelphia: Westminster Press.

MacIntyre, A. 1999. *Dependent Rational Animals*. London: Duckworth.

Oxford English Dictionary, 2nd ed. 1989. "Cooperativeness." Oxford: Oxford University Press.

Rasmussen, L. 1997. *Earth Community, Earth Ethics*. Maryknoll, NY: Orbis Books.

Roughgarden, J. 2009. *The Genial Gene: Deconstructing Darwinian Selfishness*. Berkeley: University of California Press.

Rubio, J.H. 2011. "Moral cooperation with evil and social ethics." *Journal of the Society of Christian Ethics*, *31*(1), 103–122.

Rubio, J.H. 2017. "Cooperation with evil reconsidered: The moral duty of resistance." *Theological Studies*, *78*(1), 96–120.

Sandler, R. 2007. *Character and Environment: A Virtue-Oriented Approach to Environmental Ethics*. New York: Columbia University Press.

Society for Ecological Restoration. n.d. "Society for ecological restoration." Accessed June 29, 2019. https://www.ser.org/.

Treanor, B. 2010. "Environmentalism and public virtue." *Journal of Agricultural and Environmental Ethics*, 23, 9–28.

Tudge, C. 2013. *Why Genes Are Not Selfish and People Are Nice: A Challenge to the Dangerous Ideas that Dominate Our Lives*. Edinburgh: Floris Books.

Van Wieren, G. 2013. *Restored to Earth: Christianity, Environmental Ethics, and Ecological Restoration*. Washington, DC: Georgetown University Press.

Wenz, P. 2005. "Synergistic environmental virtues: consumerism and human flourishing." In *Environmental Virtue Ethics*, edited by R. Sandler and P. Cafaro, 197–213. Lanham, MD: Rowman and Littlefield.

Yeager, D.M., and S. Herman. 2017. "The virtue of 'selling out': Compromise as a moral transaction." *Journal of the Society of Christian Ethics, 37*(1), 3–23.

Zagorsky, B.M., J.G. Reiter, K. Chatterjee, and M.A. Nowak. 2013. "Forgiver triumphs in alternating prisoner's dilemma." *PLoS One, 8*(12), e80814. https://doi.org/10.1371/journal.pone.0080814.

Chapter 8

Patience and Sustainability

JASON KAWALL

In considering the role of patience in achieving and maintaining sustainability, two conflicting intuitions seem likely to arise. On one hand, we might expect patience to play a key role: sustainability, after all, draws our attention to significant issues over extended timeframes. For example, the United States Environmental Protection Agency suggests that "Everything that we need for our survival and well-being depends, either directly or indirectly, on our natural environment. To pursue sustainability is to create and maintain the conditions under which humans and nature can exist in productive harmony to support present and future generations" (USEPA, n.d.). While there are many accounts of both sustainability and sustainable development, they share this common concern for the well-being of future generations. Our actions now will have significant impacts on the world that will be inherited by future humans and nonhumans. Seen in this light, patience—in enduring, persevering, and working toward highly ambitious long-term goals—seems likely to be a crucial virtue.

On the other hand, we might instead be struck by the extraordinary and immediate environmental and social challenges we face.

Jason Kawall, *Patience and Sustainability* In: *The Virtues of Sustainability*. Edited by: Jason Kawall, Oxford University Press (2021). © Oxford University Press. DOI: 10.1093/oso/9780190919818.003.0008

The summary for policymakers of a recent Intergovernmental Panel on Climate Change (IPCC) report suggests that

> Pathways limiting global warming to 1.5°C with no or limited overshoot would require rapid and far-reaching transitions in energy, land, urban and infrastructure (including transport and buildings), and industrial systems (high confidence). These systems transitions are unprecedented in terms of scale, but not necessarily in terms of speed, and imply deep emissions reductions in all sectors, a wide portfolio of mitigation options and a significant upscaling of investments in those options (medium confidence). (IPCC 2018, p. 17)

Particularly when focusing on climate change, it may seem that the last thing we need is patience; to encourage patience may seem to suggest a complacency and failure to take urgent problems seriously.

A related objection would hold that it is all too easy to encourage patience with respect to climate change for those who are affluent, in wealthy, stable regions with significant resources for adaptation and resilience in the face of environmental crises. But this reflects a very privileged position in the world—for hundreds of millions of people climate change is already having an impact on their lives, and their communities often lack the social and economic structures to be resilient in the face of extreme weather, sea level rise, and related challenges. We might hold, then, that to think of patience as a virtue when discussing climate change and sustainability could only appear plausible to those who are most sheltered from environmental disruptions.

In what follows I argue that patience is a crucial virtue for achieving sustainability, and that the concerns raised earlier are ultimately mistaken, resting on problematic understandings of both patience and our goals. I begin with an account of patience that draws on recent work by Matthew Pianalto and Nic Bommarito. I then present a broad range of circumstances in which (I argue) patience is essential to achieving and maintaining our goals with respect to sustainability. In the third section of the chapter I consider how addressing patience in contexts of sustainability may improve our understanding of patience more generally. And finally, in the last section of the chapter, I address the concern that patience is not a virtue at all when faced with immediate, seemingly overwhelming challenges to sustainability.

PATIENCE AS A VIRTUE

Patience has, of course, been recognized as a virtue across a multitude of philosophical and religious traditions. But here we will draw upon two contemporary accounts that, when combined, provide the basis for a rich understanding of patience as a virtue—an understanding that will guide us through the rest of this chapter.

Matthew Pianalto, in his recent book, *On Patience*, distinguishes four related aspects or manifestations of patience (2016, ch. 2). First, patience can be exhibited in calm, self-possessed waiting—as with a passenger who remains calm and collected, both inwardly and outwardly, while waiting for her delayed train to depart. Second, patience includes the uncomplaining endurance of burdens and suffering, *when appropriate*—it is this aspect of patience that underlies the use of the term "patient" in medical contexts, and it is fundamental to most theological understandings of the virtue. Such patience is exhibited, for example, by those who resolutely endure the pains and burdens of chronic illnesses. Third, there is an aspect of forbearance and tolerance to patience—as when we calmly respond to the angry outbursts of a friend dealing with a stressful, difficult situation, or to the tantrums of an exhausted child. Finally, Pianalto suggests that patience can be manifest in perseverance and constancy. We can exhibit patience in continuing to pursue a goal despite encountering difficulty or frustration, or in carefully completing the tedious tasks necessary to complete an important project. Consider an artisan who carefully—and patiently—hand-paints the fine details of an intricate, delicate work.

Underlying these elements, Pianalto treats patience as consisting in the acceptance of both unavoidable burdens and those avoidable burdens an agent herself deems to be wisely borne (2016, 53–54).[1] Still, we might wonder what allows the patient person to respond in these ways to such burdens, and here we can draw upon insights from Nic Bommarito (2014). Bommarito points to an underlying perspective on the world as fundamental to patience. While hard to define, Bommarito has in mind the sense of perspective at stake when we (for example) hold that a person who becomes furious over a trivial inconvenience lacks perspective, or when we find that a near-accident or the loss of a loved one has given us a new sense of perspective. Bommarito further illustrates the concept by identifying four aspects to the perspective underlying patience:

1. Perspective concerning our desires and values—including an understanding of what truly matters to us, allowing us to avoid frustration or despair when our trivial desires are left unsatisfied, and so on.
2. Perspective on ourselves as members of communities and as human beings—an understanding that other humans face similar burdens and sufferings, and that many are experiencing far worse than us.

1. More strictly, with respect to avoidable burdens, Pianalto notes two possibilities for patient acceptance: those burdens judged by the agent to be wisely borne or those *reasonably* judged by the agent to be wisely borne; the latter would constitute a more restrictive standard.

3. Perspective in understanding causes and laws of nature—we may be more patient or forbearing when we understand why our friend is being irritable (perhaps she is experiencing a family crisis), or when we understand the pressures that can lead ourselves and others to make short-sighted or foolish decisions.
4. Perspective in treating burdens and obstacles as opportunities for growth and improvement—resistance and setbacks can lead us to become stronger, to exercise patience and other virtues.

To Bommarito's plausible list, we can add two additional aspects:

5. Perspective concerning our abilities in general—recognizing when achieving a goal will be long and difficult (given our limitations), or accepting that some goals may not be realistically achievable by us.
6. Perspective concerning the limits of our cognitive abilities—in particular, recognizing that we will often face uncertainty, lack crucial information, and be put into positions where there is risk and inadequate guidance.

The possession of a perspective on the world including these elements would provide a plausible, unified grounding for the various aspects of patience identified by Pianalto. Realizing that other people face burdens at least as great as our own, that our abilities and understanding are finite, that our world is such that suffering and frustration will occur (even as we can strive to lessen such suffering), and that we are prone to get caught up in pursuing goods that we ourselves recognize as trivial in our reflective moments—a sense of perspective informed by such observations will make it viable for us to wait calmly, to endure suffering, to be tolerant of the shortcomings of ourselves and others, and to persevere in the face of obstacles. Following Bommarito, then, we can hold that a sense of perspective is a central, unifying feature that underlies and grounds the various manifestations of patience.[2]

ON THE NEED FOR PATIENCE FOR SUSTAINABILITY

With a characterization of patience now in hand, we can turn to a range of contexts in which it will play a key role in achieving and maintaining crucial sustainability-related goals.

Patience in Accepting Setbacks and Frustrations, in Enduring Suffering

The path to long-term sustainability has and inevitably will involve setbacks, frustrations, and suffering; we cannot expect unimpeded progress. Elections will be lost, court decisions will create obstacles, and powerful individuals and organizations will fight change.[3] Habitat loss will continue, and species will go extinct. Given current trends with respect to climate change, we can expect that many island nations will become uninhabitable, while for other populations, their homes, communities, and ways of life will be lost. There will be disease, death, and tremendous upheaval for tens of millions of people.

The struggle for sustainability will be difficult, and it will be all too easy for us to give in to despair, apathy, or unfocused and ineffectual anger. For those who face the most immediate effects of climate

2. This underlying perspective, in part constitutive of patience, will likely also play an essential, constitutive role for other virtues, including humility and *phronesis* (or practical wisdom). I hope to address such connections in future work.

3. See Eamonn Callan (1993, p. 539) on the need for patience in politics.

change—often those who already face poverty and environmental injustice—the coming decades will likely involve tremendous loss and extraordinary burdens. Over time we can work to achieve greater social sustainability and justice. But throughout this process, in the face of suffering and loss, patience can help us to carry on, and to achieve what we can.[4]

It should be stressed that patience does not require the meek acceptance of all suffering, or a mere passive resignation on the part of those most affected by climate change. Rather, it is a matter of coming to terms with this suffering, finding a way to carry on with dignity, and to avoid succumbing to despair. As Cheryl Hall notes (Chapter 4, this volume), the virtues require an understanding and perspective upon our emotions and feelings, to avoid being overwhelmed by them. And we can strive to address the sources of suffering where we can.

Compare the patience and resolve of a person working through physiotherapy following a major accident. There will be pain and frustration, but patience can allow such an individual—such a *patient*—to carry on, to endure, and to make progress in the face of setbacks and frustrations. And patience can allow us, too, to cope with those enduring, chronic pains that cannot be alleviated or improved.

Activists and those working to achieve social and environmental sustainability will face many defeats and disheartening losses. Even when things are going well, progress and improvement will take time. The scope of what they wish to accomplish—and the obstacles they face—may seem overwhelming. Activist fatigue and burnout are all too possible under such circumstances (see, for example, Chen and

4. Patience thus serves to enable and sustain hope; without patience we risk being overwhelmed and consumed by struggles and setbacks. Similarly, hope can help to sustain patience—it is easier to carry on and endure when we have hope of future success or improvement. See Thompson (2010) for a compelling discussion of radical hope in the context of climate change.

Gorski, 2015). Patience, and the perspective that underlies it, are crucial for individual activists to remain engaged. Patience enables them to face both internal obstacles (ones arising from their own psyches and limitations—such as potential frustration, anxiety, or despair) and those that are external (e.g., political setbacks, injustices, and environmental disasters). They require the perspective to accept that achieving many of their goals will inevitably involve ongoing struggle, but also the perspective to recognize the importance of what they seek and of finding a way to carry on.

Patience in Pursuing and Maintaining Long-Term Goals

In pursuing sustainability, it is important to distinguish and embrace long-term goals for ourselves and for our communities, while recognizing more generally that we have multiple goals at multiple scales. A failure to recognize the relevant differences can lead to impatience, on one hand, or complacency and lassitude, on the other.

There are many individuals with deep commitments to sustainability but who are quick to dismiss many attractive goals as "unrealistic." Thus, for example, in considering the impacts of meat-eating (and animal agriculture more generally) upon the environment, many will agree that a radical reduction or even elimination of meat-eating would likely have a wide range of positive impacts. But they worry that there is no way this could ever happen. Similarly, they might acknowledge the long-term benefits of wealthy societies becoming less consumeristic, or of increased job-sharing, or shortened work weeks, but while dismissing such proposals as merely attractive in theory, and not genuinely viable "in the real world."

However, when pressed, it often becomes clear that such individuals are thinking in terms of a relatively short time span—perhaps the next five or ten years. And certainly they are right to be skeptical of achieving radical change in too short a period. But we must keep our eyes on longer-term goals, even while recognizing our short-term and medium-term goals. Broad cultural shifts in consumption, work life, and dietary habits are quite viable, but will likely require several decades. If we are truly seeking sustainability, we should be laying foundations (in terms of infrastructure, cultural norms and values, policy, and so on) that will enable ongoing sustainability beyond the immediate need to address climate change.

This is not to deny that we also need to work on pressing changes as soon as possible to reduce CO_2 and other greenhouse gas emissions. But we must not forget that we have a wide range of goals with respect to sustainability, and that climate change is not our only challenge. Some goals may require immediate action, others may take decades or more to achieve, and there will be many others with timeframes between these extremes. Similarly, some of our goals might focus on our own personal habits, others on our local communities, while others may be aimed at generating change at regional, national, or international scales. The key point is that patience and its underlying perspective can help us to put our various goals into context, to accept reasonable timeframes for them, and to acknowledge our limitations and the setbacks we might face in pursuing them. We must avoid a narrow, impatient mindset that leads us to ignore or downplay important, viable goals that could bring benefits for decades and centuries to come.

Chapter 2 by Corral-Verdugo, Frías-Armenta and Ortiz-Valdez in this volume provides guidance relevant to embracing and achieving long-term goals. They note that stable, predictable environments can help to encourage other-regarding behaviors. Trust and sacrifice are reasonable in such contexts, while being potentially dangerous or detrimental in unstable situations. As such, working toward socially and environmentally sustainable communities can help to generate circumstances in which individuals and communities can embrace ambitious long-term goals—goals that typically require sacrifice and trust in others to be achieved (see also Hartman, Chapter 7, this volume).

Patience in Not Unreflectively Accepting Putative Solutions to Problems

A related context where patience is required lies in not being too quick to embrace supposed "solutions" to major problems without adequate attention to alternatives, or the potential longer-term impacts of such fixes. An obvious example here might be the use of highly speculative geoengineering technologies (such as stratospheric aerosol spraying to block sunlight) in an attempt to alleviate climate change. In the face of such complex, potentially catastrophic problems as climate change, activists might be tempted to embrace most any putative solution, at most any cost, and as soon as possible. Anxiety and impatience can overwhelm us when addressing large-scale, complex problems precisely because they are so intimidating. And while perhaps we will, in time, be forced into embracing such technologies, to focus on them now to the exclusion of other options seems misguided.

Patience serves as a corrective to such impulses, and encourages us to more fully consider the implications of our decisions. This does not mean waiting forever, and patience is entirely compatible with rapid action when necessary. Compare an athlete who can display patience in waiting long enough for a play to develop, allowing her teammates to escape defenders—but not waiting so long that the play breaks down. Patience can manifest itself in actions at varying scales and timeframes, from those taking mere seconds to long-term projects that last decades. The nervous athlete panics and feels compelled to try to make a play right away, perhaps as an opposing player bears down on her. The patient athlete in the same circumstances extends the play, giving her and her teammates a few crucial additional seconds for the situation to unfold and a better option to emerge.

Relatedly, studies have found that individuals who are patient in the sense of caring about future goods (and who are less prone to significantly discount such goods) are more willing to cooperate, to contribute to shared goods, and to reap the rewards of reciprocal altruism (Curry et al., 2008). Patient people may also exhibit a greater willingness to make *appropriate* compromises—given their perspective and ability to sacrifice for the sake of greater future benefits or more valuable goods. More generally, we should not give in to anxiety or impatience, and simply rush into the first available technology, weakest compromise legislation, or easiest option at a given moment. Patience encourages us to assess our options within an appropriate time frame for a given project, without sliding into complacency, on one hand (and missing crucial opportunities), or into impatience (and poor decisions), on the other.

Patience in Learning and Seeking Understanding

Plausibly, as individuals, we ought to invest time in ongoing learning about sustainability and its implications for our own behavior: who to vote for, what policies to endorse, what personal habits to change, and so on (see also Vanderheiden on conscientiousness, Chapter 9, this volume).[5] Such ongoing investigation can be draining and frustrating as we face our own lack of expertise in dealing with uncertainty, incomplete evidence, new information, and conflicting sources. We will often be tempted to assume that we are better informed than we actually are. As W.K. Clifford notes, "[w]e feel much happier and more secure when we *think* we know precisely

5. Of course, the extent of such research will vary—more would be expected of those with extensive time and financial resources than those living in poverty, working multiple jobs, and so on.

what to do, no matter what happens, than when we have lost our way and do not know where to turn" (Clifford 1999, p. 75, emphasis added). Acknowledging our own limitations and lack of adequate understanding is uncomfortable; patience will play a key role in persevering and coping with our frustrations and uncertainty.

More generally, researchers will require patience in their studies, and society as a whole will require patience in such areas as assessing policies and supporting basic research across a wide range of disciplines. It may take decades for the discoveries of basic research to influence technology or have other applications, and we risk losing out on crucial advances if we focus excessively on research directly aimed at immediate impacts.

Patient, persistent effort will be crucial to creativity, open-mindedness, and other intellectual virtues relevant to sustainability (see Wright, Chapter 10, this volume). While creativity can manifest itself in flashes of insight, more often it requires ongoing work and reflection. And when presented with new opportunities and technologies, patience will be required to open-mindedly assess them and to overcome our potential initial resistance to new ideas and ways of living. It can take time to come to grips with the novel and unfamiliar.

Similar considerations apply to education for sustainability more generally. As Matt Ferkany notes, we cannot expect single, isolated trips to a zoo with a few pious words concerning endangered species to produce genuine understanding and concern on the part of students (see Ferkany, Chapter 3, this volume). Persistence is required, with ongoing exposure to relevant information and environments enabling students to cultivate a thorough understanding of sustainability and related concerns.

Finally, we need patience in coming to understand those who disagree with us. In some instances, the gulf between us may be too great for productive engagement. But in many others, patience will pay off. We may not change the mind of a climate-change denier over the course of a conversation or two, but patience in understanding his motivations and concerns, and a willingness to engage can lead to greater understanding for both parties. We may find common ground with respect to key sustainability goals—from helping farmers adapt to changing conditions, to enabling communities to be resilient in the face of more frequent flooding or heatwaves—even if climate change deniers will deny that these changes are due to human activities (rather than "natural" fluctuations in the Earth's climate, etc.). Recognizing that individuals have a variety of goals and interests can allow for more fruitful attempts to find common ground, and to challenge and address misguided beliefs. Perseverance and a gradual, patient development of trust through ongoing interactions will often be essential to this process (see also Jain's discussion of *anekāntavāda* or pluralism, Chapter 5, this volume).

Patience in Changing Habits and Our Daily Life

To live sustainably will require a wide range of changes in our daily lives and habits—particularly for those of us among the globally wealthy. We will presumably need to become more politically active and informed, choose more sustainable housing, reduce and rethink much of our material consumption, and so on. Note that this is not to emphasize individual sustainability efforts over legislative, regulatory, and other collective efforts. As Steve Vanderheiden notes (Chapter 9, this volume), "Conscientiously seeking to reduce one's own emissions should therefore not be equated with disengagement or withdrawal from social decarbonization efforts, and it is better understood as providing the cognitive and affective bases for more effective collective action."

Among the globally wealthy, many of the required changes in habits could well be highly attractive: we might enjoy eating locally grown foods, while cycling to work could become a welcome form of exercise and opportunity to unwind. But as Cheryl Hall (2013) notes, other changes will be far less welcome—flying less (or not at all), having fewer children than we might originally have planned (or none at all), driving less (or not at all), and so forth. Achieving sustainability will not always be easy, and new habits will often be difficult to inculcate—even those that are appealing to us; patience will be invaluable in developing and sustaining new habits. Especially early on as we attempt to cultivate new habits, when we feel the impacts of such changes most profoundly, impatience and frustration will pose serious dangers. We will be tempted to slide back into comfortable old habits, to vote for politicians who will restore or maintain the familiar status quo, and to be drawn to the reassuring claims of anti-environmentalists. Having both the perspective to realize that change is often difficult for humans and the perseverance to endure and carry on despite such difficulty will be crucial to avoiding backsliding.

Over time even some of the initially unwelcome changes will be seen in a more positive light as part of a new, sustainable way of life with its own rewards. But the required work will be ongoing—sustainability is not static. We will require flexibility and patience to embrace change effectively, and the uncertainty and frustration it can bring (see also Cuomo, Chapter 6, this volume). Technologies, policies, social conditions, and countless other factors change over time. Patience enables us to persevere as we learn new skills and modify our habits and behaviors as the demands of a sustainable lifestyle change.[6]

6. Needless to say, the changes in ways of life imposed upon those facing environmental and social injustice will often be involuntary, horrific, and grossly unfair—consider the plight of climate refugees and others. But even here, the patience to endure and carry on in the face of suffering and injustice will be invaluable to those facing such circumstances.

Patience in Cultivating Other Virtues

As with the pursuit of learning and understanding, and the cultivation of new habits and ways of life, patience will play a key role in cultivating further virtues, and our character more generally. Matthew Pianalto draws attention to Gregory the Great's claim that "patience is the root and guardian of all the virtues" (Pianalto 2016, xiii). Pianalto elaborates upon the claim, suggesting that

> the lack of patience constitutes a major threat to the virtuous life, insofar as an inability to bear hardship with patience makes it immensely difficult, if not impossible, to manifest the virtues of justice, courage, and love [. . .]
>
> The virtuous action that others can witness—fair dealing with others, taking a noble risk, showing loving care to others even when it is difficult—is underwritten by a capacity to endure patiently the difficulties that arise in performing such actions. (Pianalto 2016, p. 59)

Moral growth and character development will, of course, take time and ongoing effort. In addition, as Pianalto notes, patience will often be required to allow for specific acts of virtue. Our impatience can undermine our compassion as we become frustrated with a friend struggling with difficult challenges. We may fail to act rightly when we lack the perspective and perseverance needed to address an injustice.

There is typically an important reciprocity between cultivating and acting upon patience and cultivating and acting upon other virtues. Consider a person who has cultivated a deep compassion, such that she genuinely cares about the well-being of other sentient beings (or who has embraced the virtue of *ahimsā* or nonviolence; see Jain, Chapter 5, this volume). It will be easier for her to endure frustrations and persevere—to be patient—in aiding or avoiding causing harm to other sentient beings when she cares so deeply about them. The insight and perspective she develops under such circumstances can help her to be patient in other cases, ones where she is perhaps less initially invested or concerned. Patience can thus assist us as we work to cultivate other virtues, and our cultivation of these virtues can, in turn, facilitate both the cultivation and specific acts of patience.

Of course, there are many other factors that help to enable the inculcation of patience and other virtues of sustainability. Strong, stable social environments can encourage and enable us to develop cooperativeness, trust, patience, and other virtues (see Corral-Verdugo et al., Chapter 2, this volume). Time spent in natural environments is a key factor in encouraging strong environmental identities, and effective education for sustainability—particularly with respect to ecology—can inspire an appropriate sense of perspective and of our place in nature (see the chapters by Clayton, Ferkany, and Vanderheiden, Chapters 1, 3, and 9, this volume). In turn, such a perspective can encourage wonder, awe, reverence, and other virtues and virtuous emotions (see Hall, Chapter 4, this volume).

Patience in Accepting Shortcomings in Ourselves and Others

We will often fail. We may backtrack and lose ground on habits we have been attempting to cultivate. We may lapse into apathy or despair for days, months, or even years. We will make significant decisions and perform actions that we know full well are unsustainable and for minimal benefits. We will let ourselves and others down, and we will succumb to temptations and make mistakes.

To continue on under such circumstances, despite our lapses and shortcomings, we will need to be patient with ourselves. The sense of perspective at the heart of patience will be crucial here, in recognizing our all-too-human limitations and finitude. This is not to lower our expectations of ourselves, nor to be overly indulgent. But it is to recognize that the pursuit of sustainability will be ongoing and difficult, and that little is gained if we cripple ourselves with self-doubt and recriminations. We need to find ways to carry on and to strive to do better, even when we realize that we have let ourselves down.

Our friends and allies, too, will make mistakes; our political leaders and fellow citizens will make decisions and perform actions that frustrate and disappoint us. As Joseph Kupfer notes, "Patience with the intractability of human weakness or vice marks a more advanced stage of moral maturity than the patience required, say, to wait calmly in line when people ahead of us are dawdling" (2007, p. 276). Patient forbearance and endurance will enable us to continue to cooperate with others, to rebuild trust, and to work toward shared goals. The underlying perspective of patience will help us to forgive others, allowing future cooperation (see Hartman, Chapter 7, this volume). Achieving sustainability will require ongoing work with our neighbors and fellow citizens, and forging new alliances, even if we have been let down in the past; we will need to work together.

HOW THINKING ABOUT SUSTAINABILITY CAN INFORM OUR UNDERSTANDING OF PATIENCE

Having addressed a range of circumstances in which patience will play a key role in pursuing sustainability, we can now consider how examining patience in such contexts can help to inform and shape our understanding of patience more generally. Two primary insights seem to emerge.

First, in addressing sustainability we are often concerned with long-lasting, enduring goals that will last well beyond our own lifetimes. Patience is sometimes operationalized by social scientists in terms of deferred gratification, such that patience requires only that an agent be willing to endure costs over a period of time for the sake of a greater reward later. For example, Fowler and Kam (2006) found that patient individuals, understood as those less prone to significant future-discounting, were more likely to vote. Such individuals were more willing to accept the short-term costs of becoming informed and voting for the sake of potential longer-term benefits. But in considering contexts of sustainability we recognize that the gratification at stake need not be the agent's own; often the primary benefits will accrue to others in the future.

Patience is thus not merely a virtue of personal prudence, allowing us to maximize our own flourishing by deferring the gratification of our desires. Instead, patience can promote the well-being of others across generations. Such patience has been exhibited in the past by those who have worked to establish colleges and universities, and those who have championed major social reforms—ones where the expected benefits would take decades or more to be fully realized. Sustainability brings such long-term, other-regarding patience to the fore. In choosing to act on behalf of future generations, a sense of perspective and perseverance will likely be essential, along with the patience to endure the frustrations and burdens that accompany the pursuit of such far-sighted projects.

Second, contexts of sustainability draw our attention to the importance of coping with uncertainty and incomplete information; here we can distinguish two related issues. First, there is the challenge of remaining motivated to persevere in a project even when we lack clear feedback on our progress, and when the success of the outcome is quite uncertain. Second, there is the issue of determining how long to wait or persist before abandoning a project.

With respect to the first issue, we often have projects that require sustained effort over time, and we can be quite unsure whether they will succeed. Consider an individual who becomes politically involved as a result of concerns about climate change and other sustainability issues. She might wonder whether her actions are having any impact at all—is she really helping to change minds, or to pressure her elected representatives to support green legislation? It will often be difficult for her to determine whether she is making a difference, with minimal or no clear feedback for months or even years at a time. In turn, this extended uncertainty may lead her to question herself, the value of her projects, and to wonder whether it is worth the effort; the uncertainty is troubling and can lead to impatience.

Uncertainty may also arise when we find apparently conflicting or incomplete guidance as to how best to pursue a project. It may seem that we are faced with endless choices, each bringing conflicting expert advice, or difficult-to-assess pros and cons. Those who are wealthy may be tempted to buy a new electric car, but then worry about the toxic, damaging processes involved in manufacturing the needed batteries or their implicit failure to support public transportation. We can feel overwhelmed in the face of significant issues where we lack clear, straightforward information and guidance. We may be tempted to simply give up, or half-heartedly claim that "it's all the same, it makes no difference." Or as noted earlier, we may instead be tempted to assume we know more than we do, to feel ourselves in control and well-informed.

Still, there are often aids to help us remain patient and to persevere in the face of uncertainty. (1) Even when there is much that is unknown and little feedback, we can typically receive at least *some* helpful informational input, and we can act to generate more. For example, governments might incorporate assessments of ecological footprints and other measures relevant to sustainability in guiding policy and legislation, and encourage attention and action at a variety of levels to reduce these footprints among their citizens (see Vanderheiden, Chapter 9, this volume, for related discussion). By increasing access to methods of recognizing and assessing our behavior, and by providing clear guidelines and recommendations for improvement, it will be easier for us to remain motivated and to embrace sustainable behaviors. (2) Relatedly, even as we pursue longer-term goals or enduring practices, there are often more immediate achievements along the way that can help to sustain us in our efforts. We might celebrate the return of particular species or rebounding populations as we work to restore a heavily polluted lake or river—even if the project as a whole may take decades to complete. (3) We can come to value many sustainable behaviors and habits as such, for their own sake. That is, regardless of our goals, in many cases we can come to value the processes involved, despite the frustrations and uncertainty that may arise. Compare a novelist who is often frustrated as she struggles to write, but who still values the process of writing and could not imagine her life without it. (4) In the end, though, we will still need to become adept at enduring challenging circumstances where we lack clear guidance and feedback—there will always be such circumstances, given our epistemic limitations.

The second issue concerns knowing when to stop, change course, or adjust our methods in pursuing a project. Patience typically encourages enduring and persevering, but not without limits. At some point we may need to accept that a project is not viable, or not worth the costs of further pursuit. The worry is that when deciding whether we ought to move on from a project, or change how we are pursuing it, we often lack the information needed to make an epistemically justified decision. As a simple example: do we continue waiting for a late bus when we have an important meeting to attend, or do we instead try to find an alternative? Should we keep engaging in dialogue with a person with radically different political beliefs (perhaps one who has been led to believe climate change is a hoax)—or have we reached a stage where we are simply draining time and emotional energy away from more productive ventures?

In such cases, the virtuously patient person will be sensitive to relevant cues, have some general practical knowledge (e.g., concerning how long various processes typically take), and a sense of perspective and wisdom. The patient person will thus have at least some guidance in deciding whether to stay the course. But in the end, there will be many cases where even the virtuous person will lack sufficient evidence to make a justified decision—sometimes we wait for the late bus and it arrives, and sometimes it does not.[7] This need not reflect a flaw in our patience; rather, it reflects our status as finite beings with limited epistemic abilities.

Patience is thus not a matter of waiting or persevering exactly the right length of time; humans do not possess the epistemic capacities to achieve this. Rather, patience will consist in self-possession in waiting and persevering, and in making the best epistemic judgments we can, given the overall constraints and limitations we face.[8] The virtuous person will tend to be more reliable in such judgments than the rest of us, but she will not be infallible.

Finally, we can exhibit patience in our reactions to our failures in such contexts. We may stay on a phoneline for nearly an hour, waiting to speak to a state representative's office prior to an important vote, only to discover that the call has been disconnected at some point. Here we might ruminate and become terribly frustrated with ourselves for having waited too long (or not long enough, in other cases). More generally, we might discover that we have pursued a project too long, when it was no longer viable. A key aspect of patience is an

7. There will also be cases where our judgment is epistemically well-justified, but still turns out to be mistaken.

8. As such, the patience of a human or mortal being will be more demanding than the patience of a god or ideal observer who knows exactly whether and when various events will occur, whether our projects will be successful, and so forth. Human patience involves coping with anxiety, frustration, and errors that arise precisely because of incomplete information and uncertainty—conditions that a god will presumably not face (at least given certain common understandings of omniscience).

acceptance that such errors will occur, that as finite beings our judgments will occasionally go astray.

ADDRESSING OBJECTIONS TO PATIENCE AS A VIRTUE OF SUSTAINABILITY

We can now return to the worry with which we began. Given the extraordinary pressures we face in pursuing sustainability, particularly the urgent need to address climate change, is patience a luxury we cannot afford? Is it not precisely climate change deniers and those most privileged by current social and economic conditions who would (disingenuously) encourage us to be patient, claiming we should wait before taking significant action? We can now bring together several considerations in response:

1. As we have seen, patience does not require endless, passive waiting, nor an unthinking acceptance of all burdens. We can manifest patience at a range of timescales, and through active perseverance and endurance in action; patience is a far richer virtue than the objection assumes. It is entirely compatible with patience to engage in urgent, pressing action to address climate change and other issues. Eamonn Callan nicely captures this point: "It is only a puerile, coarse-grained patience that could motivate a blanket impassivity towards evils that are fit objects of indignant resistance" (1993, p. 538).
2. Not only is patience compatible with ongoing, sustained action; it will often be necessary. Even when we engage in urgent actions as individuals (and pressure governments, corporations, and other bodies to do so as well), we will require patience: in persevering when we face defeats or setbacks, in taking the time to learn and cultivate new habits, in

forbearance when interacting with those who do not share our beliefs, and so forth. Again, the need for the endurance and perspective characteristic of patience may be especially acute for those who face the greatest burdens and suffering from the impacts of climate change and various environmental, economic, and social injustices. Even under the best of circumstances, securing and implementing major infrastructure, policy, and other changes will take time. Without patience we may be all too prone to give in to despair, change course unnecessarily, or lapse into burnout.

3. We must also remember that we are working at multiple scales as we pursue sustainability. Some goals reach far into the future—many decades or more, as when we pursue major cultural and social changes. But other goals are typically far more immediate—perhaps improving the public transportation options to a disadvantaged community, or securing adequate habitat and local cooperation to effectively protect an endangered species. Problems occur when we lose perspective on the scale and timeframes required to achieve our goals—as when we expect to create long-term changes in deeply held social attitudes over the course of five years. The latter will likely be a decades-long process, and we are bound to fail, be frustrated, or make foolish decisions if we expect to achieve such change in a short period of time. On the other hand, to take many years to properly address a contaminated water supply to a disadvantaged community in a wealthy nation is to demonstrate a vicious lack of urgency. Central to wisdom—and to patience—is recognizing the differences in scales and timeframes that are appropriate for different goals, and to have an appropriate perspective on the implications of these differences for our pursuit of each.
4. Finally, with respect to the claim that patience would require waiting for more information before acting to address climate

> change, notice that this ignores the dynamic aspects of patience and its underlying perspective; we can actively—and patiently—pursue important goals even while gathering more information. Moreover, those who would encourage inaction often seem to rely on dubious epistemic demands. Axel Gelfert (2013) describes a vicious disposition to shift epistemic standards to reduce the dissonance between an individual's background beliefs (and values) and problematic new evidence. Thus, those in the grips of climate-change denial might demand finely grained predictions of near-perfect accuracy from the models of climate scientists, while happily embracing the flimsiest of evidence that happens to support their prior political or economic views. This is not patience—it is motivated reasoning in the service of complacency.

The pursuit of sustainability, by its very nature, brings with it a need for patience—it requires the pursuit of ambitious, daunting goals across a broad range of spatial and temporal scales. Frustration, failure, suffering, and other burdens will be inevitable and ongoing. The enduring inner strength, perseverance, and perspective characteristic of patience—an often-neglected virtue—will be invaluable under such circumstances.[9]

REFERENCES

Bommarito, N. 2014. "Patience and perspective." *Philosophy East and West*, *64*(2), 269–286.

Callan, E. 1993. "Patience and courage." *Philosophy*, *68*(266), 523–539.

9. I would like to thank Heather Battaly, Jason Baehr, Alessandra Fussi, Thidwick Lorignal, and an anonymous reviewer for helpful comments and discussion.

Chen, C.W., and P.C. Gorski. 2015. "Burnout in social justice and human rights activists: Symptoms, causes and implications." *Journal of Human Rights Practice*, 7(3), 366–390.

Clifford, W.K. 1999 (1877). "The ethics of belief." In W. K. Clifford, *The Ethics of Belief and Other Essays*; introduction by Timothy J. Madigan, 70–96. Amherst, NY: Prometheus Books.

Curry, O.S., M.E. Price, and J.G. Price. 2008. "Patience is a virtue: Cooperative people have lower discount rates." *Personality and Individual Differences*, 44(3), 778–785.

Fowler, J.H., and C.D. Kam. 2006. "Patience as a political virtue: Delayed gratification and turnout." *Political Behavior*, 28(2), 113–128.

Gelfert, A. 2013. "Climate scepticism, epistemic dissonance, and the ethics of uncertainty." *Philosophy and Public Issues—Filosofia E Questioni Pubbliche*, 3(1), 167–208.

Hall, C. 2013. "What will it mean to be green? Envisioning possibilities without dismissing loss." *Ethics, Policy & Environment*, 16(2), 125–141.

IPCC (Intergovernmental Panel on Climate Change). 2018. "Summary for policymakers." In *Global Warming of 1.5°C. An IPCC Special Report on the Impacts of Global Warming of 1.5°C Above Pre-Industrial Levels and Related Global Greenhouse Gas Emission Pathways, in the Context of Strengthening the Global Response to the Threat of Climate Change, Sustainable Development, and Efforts to Eradicate Poverty*, edited by Masson-Delmotte, V., P. Zhai, H.-O. Pörtner, D. Roberts, J. Skea, P.R. Shukla, A. Pirani, W. Moufouma-Okia, C. Péan, R. Pidcock, S. Connors, J.B.R. Matthews, Y. Chen, X. Zhou, M.I. Gomis, E. Lonnoy, T. Maycock, M. Tignor, and T. Waterfield, 3–24. Geneva, Switzerland: World Meteorological Organization.

Kupfer, J.H. 2007. "When waiting is weightless: The virtue of patience." *Journal of Value Inquiry*, 41, 265–280.

Pianalto, M. 2016. *On Patience: Reclaiming a Foundational Virtue*. Lanham, MD: Lexington Books.

Thompson, A. 2010. "Radical hope for living well in a warmer world." *Journal of Agricultural and Environmental Ethics*, 23 (1–2), 43–55.

USEPA (United States Environmental Protection Agency). n.d. "Learn about sustainability." Accessed April 20, 2020. https://www.epa.gov/sustainability/learn-about-sustainability#what.

Chapter 9

Treading Lightly

Conscientiousness, Information, and Ecological Impacts

STEVE VANDERHEIDEN

How ought we to live, given the evidence that several of the impacts of our present lifestyles are ecologically unsustainable, threatening other humans and nonhumans alike? One way of answering such a question invokes voluntary or compulsory austerity, calling upon persons to sacrifice their own welfare or interests in order to live sustainably. The good life or good society is in this narrative contrasted with the sustainable one, with altruism or justice recommending the latter in what is sometimes cast as a self-abnegating subjugation of our own flourishing for the sake of others. Transition to a sustainable society, such a view suggests, must be led by moral saints, eventually to require supererogatory acts by the rest of us.

Environmental virtue theory offers a different and more attractive narrative. The good human life and good society can be oriented around a set of virtues that, once inculcated and reflected in our everyday practice, join the sustainable with the good, whether at the level of the individual person or society. Drawing upon classic

Steve Vanderheiden, *Treading Lightly* In: *The Virtues of Sustainability*. Edited by: Jason Kawall, Oxford University Press (2021). © Oxford University Press. DOI: 10.1093/oso/9780190919818.003.0009 and contemporary texts in virtue ethics, and directing moral evaluation toward character traits rather than specific actions, this approach promises an ethical outlook through which the demands of sustainability might be reconciled with desires for personal and social flourishing, and where our connections to nature might be enhanced by our cultivation of dispositions that make us more receptive to its pleasures.

The feasibility of this project depends in part upon the development of an exemplary set of such virtues, where the erstwhile opposition between the good of the individual person and the sustainability of their lifestyle is dissolved. The identification of a set of virtues that are simultaneously good for their possessor and good for the planet promises to recast the demands of sustainability as intrinsically rewarding and even essential components of a good life, rather than a necessary evil. Environmental virtue theorists have proposed numerous such candidates for this role, many of which are discussed elsewhere in this volume. Among the virtues through which this gap between the environmental right and good might be narrowed is one that I shall call *conscientiousness,* which calls upon us as persons to cultivate certain sensibilities but also requires the dissemination of several concepts and development of some social mechanisms to assist in its inculcation, allowing it to more effectively motivate and guide our actions.

FOOTPRINTS AND ATTITUDES

For humans to become more conscientious about reducing their impacts upon the planet, or for this to be beneficial, those impacts must now be sufficient to present a kind of problem. Such an assessment would be difficult to avoid when presented with data about those impacts, such as those compiled in terms of ecological footprints. Such footprints seek to quantify individual or collective human impacts in terms of the amount of land needed to support their consumption patterns through yields of ecological goods and services, including assimilation of wastes from consumption patterns.[1] According to the Ecological Footprint network, the average American requires 17.2 acres of ecologically productive land to support their consumption patterns, with the United States accounting for 12% of humanity's total footprint despite comprising less than 5% of its population (Global Footprint Network, 2015). With its relatively abundant 9.3 acres per capita of territorial biocapacity, the average American nonetheless requires almost 8 acres of ecologically productive land outside the nation's territory to supply natural resources and assimilate waste, making the United States an ecological debtor nation despite this relative abundance.

Comparing supply of and demand for ecological goods and services through the image of a calendar, Ecological Deficit Day represents that date within a year at which a nation's annual ecological budget is exceeded. A sustainable society would make it to midnight on New Year's Eve without exhausting its annual supply of ecological goods and services, with unsustainable societies exceeding their territorial ecological limits earlier in the calendar year. In 2017, the United States marked Ecological Deficit Day on June 10, not yet at the summer solstice, adding to its ecological debt for the remainder of the year (Earth Overshoot Day, 2019). Such data suggest a serious problem about which those residing in high-footprint countries or

1. Ecological footprints identify unsustainable patterns of activity in the aggregate through comparisons between the footprint of a group and surrounding biocapacity, but they can also inform normative assessments of persons or groups through carbon budgets. It is one thing, for example, to observe that humanity is consuming nearly twice the sustainable flow of ecological goods and services from our planet's biocapacity, but another to observe that my personal carbon footprint is nearly five times that of Earth's average per capita biocapacity. The first identifies a sustainability problem as the planetary scale, while the second identifies an equity problem at the individual scale (see Vanderheiden, 2008).

leading high-footprint lifestyles should be (but evidently are not sufficiently so) more conscientious about, as continued use of ecological goods and services beyond the sustainable capacity to generate them bodes ill for their future.

Indices of individual or collective impact like the ecological footprint, as with embedded carbon and virtual water, offer powerful sources of normativity from what otherwise involves a set of empirical assessments. They inform us about the otherwise-invisible byproducts of our consumption choices and alternatives, allowing for a more meaningful comparison among those options along with a powerful heuristic for critical reflection about the impacts of consumption and about the fairness of the resource allocations that make it possible. The normative premise needed to bridge the observation that four Earths would be needed to sustain all humans at my ecological footprint and the judgment that this represents an inequitable claim upon planetary resources should not be controversial (Center for Sustainable Economy, 2019). Footprint indices that measure ecological impacts and situate them within a context of what is sustainable within systems at various scales prompt the virtue of conscientiousness in that they call for attentiveness to and concern about such impacts.

Yet what would it take for residents of an affluent country with such a large per capita ecological footprint as the United States to transition to more sustainable consumption patterns, and what role does virtue play in this transition? To be sure, a sustainable transition is likely to require multiple and overlapping strategies, including public investments in upgraded energy and transportation infrastructure, policies that price pollution externalities rather than allowing waste products like carbon dioxide to be freely emitted, and changes in manufacturing and industrial supply chains to further reduce material throughputs and waste. While changes to individual consumption attitudes and behavior may contribute to or reinforce these structural and policy reforms, they represent only part of a comprehensive sustainable transition strategy rather than constituting measures that could be exclusively relied upon to close the gap between sustainable and current consumption. Wider embrace of environmental virtues may assist in this effort, but it cannot by itself close the gap between current and sustainable rates of consumption.

However, changes to individual consumption attitudes may constitute an essential part of a sustainable transition. Permissive attitudes toward excessive resource use and waste emission at the individual consumer level emanate outward to manufacturing processes and supply chains through price signals. So long as consumers engage in high rates of commodity consumption or demand products manufactured or distributed through wasteful processes, firms will continue to prioritize low prices to more sustainable products or processes, with large personal footprints manifesting through those of multinational corporations that support individual consumption. Until persons adequately value low- or zero-carbon energy or transportation options over their alternatives, they will remain a niche rather than a mainstream option for most. And unless more of those same persons attach greater weight to sustainability policies in their voting behavior, politicians can safely ignore such imperatives and avoid making policies opposed by their industry donors. In all of these ways, permissive or dismissive individual attitudes reinforce unsustainable policies and practices by governments and corporations where higher levels of individual attention to and concern about sustainability could begin to challenge them.

Mechanisms by which personal attitudes and behavior could be made more receptive to sustainability values or outcomes also vary, as the foundations of social norms and attitudes are multiple and have their origins in different aspects of contemporary life. Educational efforts designed to inform and raise awareness about anthropogenic environmental impacts, critique of wasteful consumption norms and the inculcation of experiential rather than consumerist norms, and transparency programs designed to allow organizations and persons to more readily track and thus reduce footprints have all been promoted as means by which wasteful attitudes and behaviors could be discouraged and more benign or protective ones encouraged. Since virtues like conscientiousness need to be inculcated within society in order to have practical effect, a full discussion of their role in assisting with sustainable transition would require the exploration of the means of their dissemination. Here, I shall explore how it operates as a virtue, only gesturing toward several ways in which conscientiousness might be activated or mobilized.

CONSCIENTIOUSNESS AND ATTENTIVENESS

As a virtue, conscientiousness is related to an awareness of the dynamic relationships between one's own well-being and that of the larger social and ecological community. In this sense, it can be enhanced by knowledge about ecological interdependence, just as Aldo Leopold's land ethic is at least partly the product of new thinking about ecology. Understanding the services that ecosystems provide for humans, from nutrient cycling to carbon sequestration, contributes to this awareness, and thus, inculcation of the virtue is assisted by science education about ecological systems. Education in ecology can therefore provide some basis for the kind of cognitive mindset needed to embrace and act upon the virtue, but formal scientific knowledge comprises only a part of what is needed. As Matt Ferkany notes in the context of education for sustainable development, the mindset needed for inculcating environmental virtues includes capacities for critical thinking and education in social and political ideals like respect for others and commitments to democracy and peace (Ferkany, Chapter 3, this volume).

It is not merely the science of ecology that contributes to this awareness, as Leopold notes in his famous essay (1949). That many persons lack what he termed an "ecological conscience" owes also to social and economic relationships of production and the geographic and cognitive distance from the land that many such relationships can impose on our thinking. As he writes:

> Your true modern is separated from the land by many middlemen, and by innumerable physical gadgets. He has no vital relation to it; to him it is the space between cities on which crops grow. Turn him loose for a day on the land, and if the spot does not happen to be a golf links or a "scenic" area, he is bored stiff. If crops could be raised by hydroponics instead of farming, it would suit him very well. Synthetic substitutes for wood, leather, wool, and other natural land products suit him better than the originals. In short, land is something he has "outgrown." (p. 210)

The separation that Leopold laments—physical, economic, epistemic, psychological—results from social and economic arrangements by which the "true modern" person not only fails to properly appreciate the links between ecosystem health and the continued satisfaction of their material preferences and lacks any feedback mechanism for identifying changes in ecosystem health, but falsely comes to believe humanity to have "outgrown" or transcended nature itself.

The social inculcation of the virtue therefore entails attending to the sources of this view of humans having transcended nature. As Lynn White has persuasively argued, the historical roots of this human exceptionalism date to early Christian rejection of Franciscan animism, when the ontology of humans as outside of or superior to the natural world was developed, as well as with the nineteenth-century separation of science from technology, with the latter oriented around the human transcendence of natural limits (White,

1967). As White suggests, "the issue is whether a democratized world can survive its own implications," having spawned an ontology of humans as separate from and superior to the natural world (White, 1967, p. 1206). Environmental philosophers have engaged in critical work to recover a sense of human embeddedness within nature, with the identification of humility as among the environmental virtues signaling the retreat from an ontology of what Leopold cast as viewing humans as "conquerors" to viewing them as "plain citizens" of the biotic community.

In addition to ontological foundations of viewing humans as embedded within rather than apart from or conquerors of nature, and as indices like the ecological footprint illustrate, the awareness component of the virtue can be enhanced through information and transparency that quantifies and makes palpable our human impacts upon the environment. Whether in terms of the relative energy efficiency of appliances, the carbon embedded in our transit choices or water embedded in our food, or the ecological footprints of our individual lives or communities, such information can contribute to our critical self-assessment (Vanderheiden, 2016). We can get from these measurement efforts a rough idea of the ecological inputs that are needed to support our consumption activities as well as the waste outputs from them, relative to other persons as well as to what is possible through available alternatives. While such information cannot by itself make us conscientious, as this also entails a set of normative judgments about what we ought to do, it can increase our awareness of our ecological interdependence, upon which the virtue also depends.

Contesting this separation and seeking to restore some greater human connection to the natural world may have ancillary benefits that solidify the status of conscientiousness of a virtue necessary for human flourishing. Martha Nussbaum identifies "being able to live with concern for and in relation to animals and the world of nature" as among the 10 core human capabilities necessary for a flourishing life and thus an object of justice (2006, pp. 78–81). E.O. Wilson identifies this innate human desire for connections with nature to be the product of evolution, proposing the biophilia hypothesis as describing "the connections that human beings subconsciously seek with the rest of life" (1992, p. 350). Evidence suggests that biophilic design principles can help to "create good habitat for people as a biological organism in the modern built environment that advances people's health, fitness and wellbeing" (Kellert and Calabrese, 2015, p. 6). As Susan Clayton notes, experience in nature itself rather than mere contact with natural elements in the built environment assists in the formation of environmental identities, strengthening this association (Clayton, Chapter 1, this volume). All suggest that the connections with nature that we seek as aspects of human flourishing rely upon this aspect of conscientiousness, through which our links to the natural world are awakened, enhancing our well-being.

While conscientiousness is primarily attuned to minimizing our harmful impact upon the environment, its ontological and normative orientation depends also upon our appreciation of the manifold ways that humans are constituted by ecological systems and receive essential benefits from them. Because it does not rely only upon the threat that humans pose to the environment from affluent consumption patterns, but also draws upon the manifold interactions within and services derived from nature in establishing or reinforcing our place in it, conscientiousness as a virtue need not be reserved for the high-impact consumer from the global North. Those living in poverty and so more vulnerable to ecological degradation and dependent upon maintenance of ecological services can exercise the virtue through stewardship of nature informed and motivated by anthropogenic threats to its integrity rather than through a conscious desire to mitigate those threats posed by their personal consumption patterns. The virtue arises not only from recognition of nature's vulnerability to human activity, but also from human vulnerability to degraded or depleted nature. It is from this recognition of embeddedness and dependence that the prescriptive component of the virtue arises, as shall be considered next.

FROM ATTENTIVENESS TO CONSCIENTIOUSNESS

Beyond mere attentiveness to our connections with the environment we inhabit, the virtue of conscientiousness assumes a desire to do what is right. This prescriptive component contains both a negative and a positive component. Negatively, it reminds us to tread lightly upon the Earth, refraining from harm resulting from excessive impact. To be conscientious is thus to be aware of our potential negative impacts and to seek to minimize them, for the sake of others or the environment itself. But the virtue also has a positive component, discussed here by Thoreau (1854):

> I know of no more encouraging fact than the unquestionable ability of man to elevate his life by a conscious endeavor. It is something to be able to paint a particular picture, or to carve a statue, and so to make a few objects beautiful; but it is far more glorious to carve and paint the very atmosphere and medium through which we look, which morally we can do. To affect the quality of the day, that is the highest of arts. Every man is tasked to make his life, even in its details, worthy of the contemplation of his most elevated and critical hour. (p. 142)

As Thoreau suggests, the conscientious seek to improve the world through their participation in it, as exemplified in his role as participant-observer in the natural world surrounding his Walden cabin. Not that humans must transform nature or leave their imprint upon it through this kind of participation, as this can often be harmful, but Thoreau suggests a creative force in living that when directed toward the natural world has the potential to elevate both.

Perhaps the closest analogue to this positive component of environmental virtue is the commitment to the good of the *polis* in civic virtue, or through activities of citizenship. For Aristotle in *Politics*, the virtue of the good citizen is distinct from that of a good person, as the task of the citizen is the "preservation of the partnership" within the regime, where the specific content of this virtue depends upon regime type but where the civic virtues are those that incline citizens toward the common good. The classic civic virtues are thus the qualities that tend toward the maintenance of the *polis*, not merely the *telos* of the individual person, and involve positive obligations to support and maintain its institutions and civic life. This quintessentially involves participating in self-governance, where citizens join together to promote the common good and the civic virtues are those which incline persons toward the kind of engagement.

The difference between civic and environmental virtues, then, lies in their orientation toward the integrity of the environment in which the *polis* is nested, not toward any particular regime type or political community. Membership is defined in the context of ecological rather than social or political boundaries, with the relevant virtues oriented toward the advancement of the interests of this more broadly defined community. The idea of "environmental citizenship" derives from this conception, and as originally developed by Environment Canada notes that:

> Each of us has an effect on the environment every day; the key is to make this impact a positive one. We must all take responsibility for our own actions, whether as individuals, or as members of a community or an organization. (Quoted in Bell, 2005, p. 179)

As developed by Andrew Dobson through what he terms "ecological citizenship" (2006a), its community is "created by the material relations of cause and effect" rather than as defined by territorial borders or cultural distinctiveness, but "is a space in which *political* obligation operates" (p. 447). According to Dobson, our "embodiedness" in the world "gives rise to the ecological footprint" which aims to measure the extent of our personal impacts upon that world, which he understands as "a network of effects that prompts reflection on the nature of the impacts they comprise" (2006b, p. 177).

For Dobson, the ecological footprint not only informs us about our impacts upon the planet's ecological systems, but it links us to those suffering from environmental harm through a "thick cosmopolitanism" of globalized cause and effect and allows us to conceive of our just share of the earth's ecological goods and services. In this sense, it performs some of the same roles that conscientiousness entails, calling attention and providing a normative orientation for our actions. The conceptual "space" that he compares to bounded political communities of ordinary civic obligations involves an expansive human community inclusive of all those sharing a planet and its ecosystem services, where the recognition of ecological interdependence coupled with a sense of potential culpability and notion of distributive justice defines and motivates our self-imposed constraints, through which we adhere to equitably assigned ecological limits. For those inclined to view themselves as belonging to a community from which civic obligations to support that community follow, the integrated global conscientiousness that ecological footprints foster enlarges that domain beyond the nation-state.

Others, for whom cosmopolitan citizenship is too expansive or abstract, can deepen their connections with bounded communities by viewing their impacts upon and embeddedness within more local environments through the construct of the ecological footprint, albeit without the global reach that Dobson describes. Attentiveness to and conscientiousness about our reliance upon others in society for the provision and maintenance of public goods, including but not limited to ecological goods and services, reinforces the fact that society constitutes an aspect of our environments. As an activity associated with maintaining the necessary external social and environmental conditions for flourishing, citizenship can be viewed as integral to social as well as environmental sustainability. While some civic virtues emphasize the natural and others the social dimensions of environmental quality, conscientiousness can be responsive to both, even if indicators like the ecological footprint call primary attention to the former. By expanding the boundaries of community to include all of humanity, it transcends the merely social in its reach but reinforces the identification with human communities by connecting individual footprints with broader social impacts and imperatives to more equitably share conservation burdens. Importantly, it links social with environmental sustainability in terms of cause and effect.

Conscientiousness therefore asks that we be responsive to the fact of our embeddedness in the environment, recognizing our own dependence as well as the potential impacts that we might have on others. Whereas our response turns on normative judgements about what we ought to do or how we ought to live, it builds upon what Rom Coles characterizes as *receptivity*, described here as a democratic commitment but applicable also to environmental virtue (2004). "I emphasize the term receptivity, rather than listening," Coles writes, "because I think that the former term evokes a broader notion of responsiveness and helps attune us to a broader range of practices" (p. 684). Being receptive to modifying our actions out of the recognition of ecological limits and commitments to equity, as Dobson suggests, can inform a view of citizenship in which the relevant virtues orient the individual within the community and both within the environment. It is a form of listening to nature, and of being attentive to the feedback that we receive.

Given its prescriptive component, conscientiousness offers an example of what Ronald Sandler terms an "environmentally productive virtue" in that the virtue "aim[s] at protecting or promoting environmental goods and values" by reducing adverse environmental impacts and enhancing the human relationship with nature (2013, p. 4). While a companion virtue like stewardship may attend to personal opportunities for transformative environmental enhancement, whether through picking up litter left by others or through the cultivation of plants or restoration of damaged sites, conscientiousness aims primarily at damage reduction or avoidance. Its corresponding vices are thus inattentiveness, which manifests as an indifference toward personal environmental impacts, as well as profligacy, where excessive consumption of resources or production of wastes results from this indifference in combination with adherence to social norms permitting or encouraging wasteful consumption. A key aim of seeking to inculcate the virtue is to reduce the human impacts upon the environment, not merely to increase our appreciation of the role that nature plays in supporting and enhancing our welfare.

That conscientiousness aims to mitigate human impacts need not impugn its status as a productive virtue, however, as the virtue can be productive in nonmaterial ways, as well, including through the fostering of healthier relationships between humans and nature. Treading lightly, or the ability to enjoy the natural world without damaging it or depleting its resources, is an attribute of sustainable living and an environmental value. Moreover, as Sandler notes, virtues like conscientiousness allow us to relate to the natural world differently than we would if acting from their corresponding vices.

> Those who possess traits like wonder, humility, and love experience nature and relate to it differently than those who are arrogant, lazy, and indifferent to the natural world. For the former, nature often is a source of nurturing, renewal, knowledge, and

> joy; for the latter it is less likely to be so, since they are less likely to go into nature and explore, study, reflect, and appreciate it. (Sandler, 2013, p. 6)

To be conscientious about one's personal impacts upon the environment is to simultaneously manifest care for nature and concern for others, which involves an affective tie that cannot be fully captured as exhibiting only self-restraint. Being conscientious in this sense, and acting accordingly, enacts a relationship between self and other that is at least potentially productive as well as exhibiting necessary restraint against harm.

Conscientiousness is more than mere attentiveness, insofar as being attentive requires only that one cognitively acknowledge the outputs of one's actions. If I was, for example, to keep a daily log of my carbon and water footprints, not as part of a plan to minimize them but merely from a rote exercise in recording data, I would be attentive to but not conscientious about those outputs. My attention would extend to the outputs themselves, but not to their connection to the world or in a comparative sense what would be sustainable for a person like me. Likewise, if I was to collect such data for the purpose of seeking affirmation from others or only appearing to be ecologically concerned, my actions would lack the conscientiousness needed to motivate my genuine effort to live more ethically. It is only through a commitment to mitigating those impacts, and to developing a closer cognitive and affective relationship with nature through which we recognize our interdependence with others in the ecological systems that we share that attentiveness becomes conscientious. This recognition, in turn, combines with acknowledgment of the rights or interests of others to exist and flourish within our shared spaces to deepen attentiveness to our footprints such that we desire not only to reduce them and thus to live more sustainably, but also to do so on an ethical basis in the role of what Leopold termed the "plain citizen" of the biotic community. As a virtue, it commends that its knowledge base serve to direct ethical action through the cultivation of a virtuous character. Its contrast with action-oriented ethical theories shall serve as a further point of contrast.

VIRTUE AND ITS ETHICAL RIVALS

As a virtue oriented around the evaluation of character, rather than a directive for action, conscientiousness need not be outcome-oriented, as are consequentialist ethics that speak to imperatives of reducing personal carbon footprints. Such an approach may allow ethicists to sidestep several thorny theoretical matters inherent in ethical evaluation of individual actions, and it provides an alternative basis for theorizing personal commitment in the absence of political mandates or collective action. The advantages of starting with virtue and character rather than consequences and individual causation are evident in contrast with other views of climate ethics.

John Broome, for example, argues that persons have a moral obligation to attain carbon neutrality, offsetting their personal carbon footprints where necessary, in order to avoid complicity in climate-related harm. In reference to what he characterizes as the private morality of individual ethics and contrasts with the public morality of government regulatory actions, Broome argues that persons have a "duty of justice not to harm, rather than the aim of improving the world" (2012, p. 14). To achieve this end, he argues, persons can "cancel" any individual culpability in climate-related harm by purchasing voluntary carbon offsets, since "emitting a tonne of carbon dioxide and offsetting it is exactly as good as not emitting it in the first place, providing the offset is genuine" (p. 52). For Broome, both the duty and the means of discharging it are understood primarily through a consequentialist framework, with the agent seeking to avoid causing harm associated with climate change by ceasing net emissions of the greenhouse gases that cause it.

In allowing persons to avoid complicity by paying others to reduce carbon emissions rather than attending to their own carbon footprints, however, Broome not only facilitates the shifting of burdens away from wealthy polluters by allowing them to continue polluting where economically beneficial to do so, but he may also undermine the virtues association with seeking to live a more sustainable life. As Jeffrey Skopec (2010) argues,

> One can achieve carbon neutrality despite living a life of excessive consumption; it does not matter if one is offsetting the minimal carbon produced by a hybrid car, or the substantial carbon produced by a private plane. In this way, the carbon market obscures the idea that the environmental ethic against wastefulness is not just about outcomes, but about an ethos—a type of character. (p. 2086)

In contrast with a consequentialist ethic like Broome's, a focus upon virtue would encourage those character traits and seek to habituate persons into low-carbon behaviors, viewing waste and excess as the problem and orienting attitudes and dispositions against it. In this way, the focus upon character and virtue offers a more compelling case of what it means to take on personal mitigation commitments than does Broome's emphasis upon carbon neutrality, a standard that would be relatively easy for some to attain but exceedingly difficult for others, particularly with the option of reaching the target through the purchase of carbon offsets.

The focus upon character and habit also views individual mitigation actions in terms of patterns of behavior or sustained efforts rather than a series of one-off choices, and it emphasizes the ethical integrity of the agent rather than any putative benefits for action. The latter may confer more meaningful ethical guidance in cases where persons find themselves surrounded by others that neglect to take action to reduce their contributions to climate change. In such cases, it is unlikely that the unilateral actions of small numbers of concerned persons will have beneficial effects upon the climate in the face of widespread indifference by their neighbors. One may still be virtuous in the midst of widespread vice, and indeed such resistance is often more impressive when social norms permit pollution than when they do not. Remaining conscientious about one's own carbon footprint under such circumstances is therefore undiminished by the ethical failures of others, commending moral courage where it is most needed.

Calling for moral courage and leadership is more difficult through rival approaches to climate ethics. Elizabeth Cripps (2013), for example, considers whether persons in such circumstances might have "mimicking duties" for persons residing in countries that do not effectively mitigate the carbon emissions of their resident population. Seeking to confine ourselves to a fair personal carbon allowance in the midst of widespread excess and indifference might, she suggests, be an act of "moral self-indulgence" insofar as this involves exclusively pursuing an individualized mitigation effort when potentially more effective collective ones are needed. "If a great harm requires a collective solution," she argues, "it is at best disingenuous—at worst repugnant—to wash one's own hands, morally speaking, and leave the resulting catastrophe to unfold" (2013, p. 132). If environmental virtue is centrally concerned with maintaining individual moral integrity and not commitment to a larger community, such a critique has considerable force.

But conscientiousness does not seek to escape from complicity in ongoing environmental harm—indeed, it acknowledges such harm while also seeking to reduce individual contributions to it. Conscientiously seeking to reduce one's own emissions should therefore not be equated with disengagement or withdrawal from social decarbonization efforts, and it is better understood as providing the cognitive and affective bases for more effective collective action. The virtuous citizen cannot be conscientious without also being engaged in necessary efforts to avoid the kind of catastrophe Cripps identifies, while also providing motives for adherence to conscientious effort to live sustainably in the absence of ongoing collective efforts. As suggested earlier, the social inculcation of attitudes conducive to greater care for and protection of the environment is likely to support rather than conflict with the various other strategies for sustainable transition, including those operating at the collective and policy levels.

The primary distinction between the virtue-ethics approach described here and the approaches of Broome and Cripps lies not in whether or not persons take adequate responsibility for their contributions to global problems like climate change, but lies in the nature of virtue ethics itself, which focuses upon character and habituation and endorses a plurality of virtues rather than any single ethical imperative. Herein also lies the value of the virtue-ethics approach, which offers a more compelling narrative about how to live when others neglect to do so while also furnishing the motivation, attending to the information needed to beneficially direct action, and providing the sustained commitment toward a manner of living compatible with the imperatives of sustainability and our shared inhabitation of an ecologically finite planet.

CONCLUSIONS

Like social norms, virtues orient persons toward the social good, not necessarily toward a narrow or materially oriented individual self-interest. Courage in battle, for example, is typically recognized as virtuous for soldiers in that its absence would result in greater vulnerability against attack. It thus serves the collective interests of armies and of the polities that raise them to encourage courage among soldiers, but cowardice may improve the chances of a particular soldier surviving a battle. In a similar fashion, generosity toward the needy might be regarded as collectively beneficial in that those needs might be better attended thereby, but stinginess may nonetheless be individually rational. Virtues, like social norms, are socially cultivated in part to reconcile this conflict between individual and collective rationality by encouraging persons to identify with collective interests, then to apprehend an interdependence of individual and social interests and reinforce the personal value of maintaining public goods. But once inculcated and equated with individual as well as social flourishing, as with Aristotelian *eudaimonia*, the virtues can help to reconcile apparent conflicts between the interests of persons and those of their communities, while also providing the motive and direction for advancing both.

Conscientiousness regarding the impact of one's lifestyle or behavior upon the world does not obviously serve self-interested objectives, as the additional burden entailed by adopting low impacts behaviors might outweigh any marginal improvement in environmental quality that resulted from the change to a single person's ecological footprint. Unlike conscious attention about one's exposure to environmental hazards, which can benefit those persons displaying such characteristics (even if it does not always), the benefits that arise from conscientiousness accrue to whole societies or even to the entire planet. To the extent that one inculcates the virtue, it does not obviously serve individual self-interests so much as the flourishing of the collective in return. Some may view the virtue as altruistically oriented as a result, or regard the directives that it commends as entailing some kind of sacrifice of the individual to the community. Unless free riding upon the efforts of others is cast as a vice and appropriately stigmatized, this gap between the collective focus of virtues like conscientiousness and the individual costs of adhering to them might undermine their efficacy in galvanizing cooperative and sustainable behavior in reducing anthropogenic environmental threats.

Within the context of a relatively small nation-state, conscientiousness may not result in net social benefits if widely adopted, given its widely distributed and diffuse benefits. If all the residents of Australia, for example, were to become conscientious about their carbon footprints, undertaking expensive decarbonization measures in an effort to reduce them, the costs of their mitigation efforts may exceed the benefits to that 0.33% of the world's population that bore those costs. Hence, conscientiousness may be a global human virtue rather than a social one—or be based in what Dobson refers to as a "deterritorialized" conception of citizenship (2003)—at least insofar as the virtues are understood as primarily oriented toward the collective good. But for each member of the global human collective, now oriented toward one another as members of a common community, the senses of commitment and belonging that arise may not only vindicate the efforts they require but may also foster the ontology needed for humans to live sustainably.

REFERENCES

Bell, D. 2005. "Liberal environmental citizenship." *Environmental Politics*, *14*(2), 179–194.

Broome, J. 2012. *Climate Matters: Ethics in a Warming World*. New York: W.W. Norton & Co.

Center for Sustainable Economy. 2019. "Ecological footprint quiz." Accessed April 25, 2019. http://myfootprint.org/en/.

Coles, R. 2004. "Moving democracy: Industrial area foundation social movements and the political arts of listening, traveling, and tabling." *Political Theory*, 32(5), 678–705.

Cripps, E. 2013. *Climate Change and the Moral Agent.* New York: Oxford University Press.

Dobson, A. 2003. *Citizenship and the Environment.* New York: Oxford University Press.

Dobson, A. 2006a. "Ecological citizenship: A defence." *Environmental Politics, 15*(3), 447–451.

Dobson, A. 2006b. "Thick cosmopolitanism." *Political Studies, 54*(1), 165–184.

Earth Overshoot Day. 2019. "National ecological deficit days." Accessed April 25, 2019. https://www.overshootday.org/about-earth-overshoot-day/national-ecological-deficit-days/.

Global Footprint Network. 2015. *State of the State: A New Perspective on the Wealth of Our Nation.* https://www.footprintnetwork.org/content/images/article_uploads/USAFootprintReport_final_lores.pdf.

Kellert, S.R. and E.F. Calabrese. 2015. *The Practice of Biophilic Design.* http://www.biophilic-design.com.

Leopold, A. 1949. *A Sand County Almanac.* New York: Oxford University Press.

Nussbaum, M. 2006. *Frontiers of Justice: Disability, Nationality, Species Membership.* Cambridge, MA: Harvard University Press.

Sandler, R.L. 2013. "Environmental virtue ethics." In *International Encyclopedia of Ethics,* edited by H. LaFollette. Malden, MA: Blackwell. https://doi.org/10.1002/9781444367072.wbiee090.

Skopec, J.M. 2010. "Uncommon goods: On environmental virtues and voluntary carbon offsets." *Harvard Law Review, 123*(8), 2065–2087.

Thoreau, H.D. 1854. *Walden.* New York: Houghton, Mifflin & Co.

Vanderheiden, S. 2008. "Two conceptions of sustainability." *Political Studies, 56*(2), 435–455.

Vanderheiden, S. 2016. "The obligation to know: Information and the burdens of citizenship." *Ethical Theory and Moral Practice, 19*(2), 297–311.

White Jr., L. 1967. "The historical roots of our ecological crisis." *Science, 155*(3767), 1203–1207.

Wilson, E.O. 1992. *The Diversity of Life.* Cambridge, MA: Belknap Press of Harvard University Press.

Chapter 10

The Roles of Creativity and Open-Mindedness in Implementing Sustainable Solutions

SARAH WRIGHT

WHY INCLUDE INTELLECTUAL VIRTUES IN OUR ENVIRONMENTAL VIRTUE ETHICS?

When developing an environmental virtue ethic, it is natural to focus on the attitudes that we expect people to take toward the natural world and the actions that we think that they ought to refrain from or engage in. It is less obvious that we need to consider the epistemic characteristics of environmentally virtuous people. But environmental virtues clearly require their possessor to meet certain epistemic requirements. We can see this by considering four ways distinct ways that authors have picked out environmental virtues, as surveyed by Ronald Sandler (2005). He notes that different approaches to the environmental virtues will generate different ways to identify them, which he classifies into four general kinds. Each requires an epistemic component.

Sarah Wright, *The Roles of Creativity and Open-Mindedness in Implementing Sustainable Solutions* In: *The Virtues of Sustainability*. Edited by: Jason Kawall, Oxford University Press (2021). © Oxford University Press. DOI: 10.1093/oso/9780190919818.003.0010

The first way to identify environmental virtues is through an extension of traditional interpersonal virtues. Geoffrey Frasz (2005), for example, has argued that virtues like benevolence and gratitude, traditionally seen as character traits concerning our attitudes toward other people, ought to be extended to nature. Yet, in order to be truly grateful to nature, we need to know at least some of the ways that nature has benefited us. In order to be truly benevolent toward nature, we must know how to treat nature in ways that are beneficial to it. Thus, virtues picked out, by extension, will require that we seek out information about nature.

The second, more egocentric, method of identification is that environmental virtues ought to benefit their possessor. This fits with the traditional idea that virtues are required in order for us to live a good life. In the environmental case, we can see that harm we do to our environment often brings about costs and harms to ourselves. Even if we are primarily motivated to benefit ourselves, there are still many facts about nature that we need to know. How does our environment help us? What ecosystem services does it provide? How might we use its resources in a sustainable way to make our lives better? Answering each of these questions requires rigorous and careful inquiry. So epistemic virtues will be a beneficial supplement to egocentric environmental virtues.

Sandler's third way into environmental virtues asks us to look at the etymological roots of "virtue" in the Greek concept of *areté* or excellence. Human excellence requires the development of all our faculties, including the intellectual ones. While we traditionally think of virtues as focusing on excellence in our feeling and acting, our epistemic capacities are also clearly ones that can be exercised in better and worse ways, so their excellence should be part of the overall excellence of a person.

Finally, Sandler considers how we can come to understand the environmental virtues through studying notable exemplars. The examples he gives are John Muir, Rachel Carson, and Aldo Leopold. We admire these individuals for their commitment to preserving the environment, and their abilities to see the natural world in ways that others do not. We look to those elements of their personalities to give us a model of how we ought to feel and act toward the natural environment. These individuals also stand out in their epistemic abilities; we admire them for their knowledge and understanding. But we admire them not just for the fruits of their epistemic labor; their intellectual courage and carefulness also stand out. These traits of epistemic character are good candidates to be considered intellectual virtues. Thus, the exemplars that we respect embody both environmental and intellectual virtues.

Environmental virtues, then, whichever methods we use to identify them, will make epistemic demands on us. All virtues do so, in so far as we need to be practicing practical wisdom to be acting virtuously. But the information required for environmental virtues will often be far more complex than that required for those ethical virtues that concern how to treat others we deal with directly. The effects of our actions on the environment can be wide-ranging, dispersed, and long-lasting. To determine which actions are virtuous, we will need to have an extensive knowledge of ecological interactions and how we can influence them.

Jason Kawall has been concerned with specifying how much environmental virtues ought to require of us epistemically (2010). Once we recognize the enormity of our environmental problems and so realize the complexity of the information that is needed to link our actions to their environmental impacts, we may simply become paralyzed into inaction. How can we act when it is so difficult to know how our actions will affect our local and global environments? And yet we also need to make sure that we are balancing our moral and epistemic demands. An endless search for more information can stop us from acting as we ought. Kawall suggests that it is appropriate for us to winnow down this arduous task by choosing to focus our efforts in particular spheres, those we believe to be important and about which we can have an impact. He argues that we ought to seek out enough information to act in a morally justified way in only those spheres. On other questions we are forced to settle for acting only in a nonculpable way; one that is compatible with the best of our evidence, but about which we have not been able to gather sufficient evidence for justified belief.

If we are approaching environmental ethics through an environmental virtue approach, the addition of intellectual virtues is a natural way to address our competing moral and epistemic obligations. Intellectual virtues, as virtues, should be a part of an overall good life; intellectual virtues should not leave us paralyzed by our search for information. The demand that moral and intellectual virtues be jointly satisfiable puts limits on them both; one cannot act without information, but one ought not inquire to the exclusion of action. Thus taking a virtue approach to our epistemic obligations presents itself as a fruitful route.

But not all accounts of the intellectual virtues are equal on this count. Between virtue reliabilism and virtue responsibilism, only the responsibilist approach really provides us with unity between the moral and intellectual virtues. Virtue reliabilism bears a clear resemblance to reliablism in epistemology in general. Reliabilist intellectual virtues are dispositions or traits of the agent that reliably produce true beliefs. Virtue reliabilism differs from reliabilism in general by requiring that the relevant true-belief-producing dispositions be integrated into the agent; fleeting dispositions are not considered virtues. However, the range of virtues recognized by this view are still broad; both in-born abilities like eyesight and hard-won character traits like open-mindedness are candidates to be intellectual virtues (see Greco, 2010; Sosa, 1991). Reliabilist virtues have two important limitations. First, reliabilist virtues don't reflect positively on the person possessing them. Having good eyesight doesn't make someone a better person. Second, since they are not virtues of character and do not resemble moral virtues, we have no reason to think that they can be integrated with the moral virtues into an overall good life. They are simple abilities that do not come with instructions for how or when they are to be used. But we are interested in the question of when we ought to seek more information and when we ought to rest content, and this question is one that reliabilist intellectual virtues cannot answer.

In contrast to virtue reliabilism, virtue responsibilism is an approach to virtue epistemology which does focus on analogies between the moral and the intellectual virtues. While virtue responsibilism has been developed in different ways by different authors, these accounts share a focus in developed traits of intellectual character that in some way reflect on the evaluation of the person possessing them.[1] While virtue reliabilists search broadly for all the cognitively integrated dispositions of the individual which reliably produce true beliefs, responsibilists focus more narrowly on developed traits of intellectual character. This narrow focus fits with the responsibilists' idea that intellectual virtues will (like moral virtues) reflect positively on the person who has them. While good vision might be quite useful, and while it might reliably lead to true beliefs, we do not think that the person with good eyesight is a better person than someone with poor eyesight; as a result, virtue responsibilists will reject eyesight as a virtue. Open-mindedness, however, is a disposition in an individual that plausibly has to be developed and maintained over time. It also reflects on the evaluation of the person possessing it; an open-minded person is a better person than someone who is closed-minded. Open-mindedness, intellectual courage, and intellectual

1. Central examples of responsibilism include Zagzebski (1996); Code (1987); Roberts and Wood (2007); and Baehr (2011).

humility are among the character traits that have been considered as intellectual virtues by responsibilists.

Virtue-responsibilist intellectual virtues are also poised to be action guiding in a way that reliabilist virtues are not. The reliabilist framework can tell us which processes and methods to avoid using—those that bring us to false belief more often than true belief. But they cannot tell us further when, how, and to what degree we ought to use reliable processes. Responsibilist virtues aim to do this.

Finally, the responsibilist approach to intellectual virtues aims for integration between the moral and intellectual virtues. They should both be exemplified in a good life, and so their demands cannot be in direct conflict. In cases where the moral and intellectual virtues appear to be in conflict, just as in cases where different moral virtues appear to be in conflict,[2] we will need practical wisdom to resolve the apparent tension between them, guiding us to the correct course of action between competing demands.

Since responsibilist intellectual virtues are the best fit to supplement environmental virtue ethics, we will look for virtues that fit this model as we explore which intellectual virtues are needed for the implementation of sustainable solutions.

FOCUSING ON SUSTAINABLE SOLUTIONS

The World Summit on Sustainable Development in 2002 established three pillars of sustainable development: economic, social, and environmental. The three pillars supplement their articulation of the goals of sustainable development: "Humanity has the ability to make development sustainable—to ensure that it meets the needs of the

2. Note that the introduction of environmental virtues will also make the possibility of apparent conflict between the virtues more likely.

present without compromising the ability of future generations to meet their own needs." Thus sustainable development aims to promote all of these pillars, not willingly making a large sacrifice in one category just for gain in the others; in particular, we should not privilege current social and economic needs over those of the future by sacrificing ecosystem health for economic development.

My focus here will be on what I will call "sustainable solutions." These are projects that make up a part of overall sustainable development. Sustainable solutions should adhere to the goals of sustainable development by offering economic, social, and environmental net benefits. That is, an ideal sustainable solution should benefit all three pillars of sustainable development; we are looking for those (perhaps rare) projects that take advantage of an opportunity to benefit all three values collectively. I focus on the most ideal of sustainable solutions because even their discovery and implementation requires the exercise of the virtues I develop here. Projects that involve making a trade-off between these three pillars will clearly involve even more difficult choices about whether and how they should be implemented. But my focal virtues will still be required even in easy choices.

These sustainable solutions will be individual projects, but the overall goal of sustainable development will be to build these smaller projects up into systems that achieve the same end. I focus here on the smaller projects in order to more clearly identify the intellectual virtues that might be required for their implementation. But I anticipate that the same intellectual virtues will be relevant at all levels of sustainable development.

AN EXAMPLE OF A SUSTAINABLE SOLUTION

Exploring a concrete example of a sustainable solution will allow us to examine which intellectual virtues are most relevant to its discovery and implementation. The example that I will use is a recent project in Kenya that turns human waste into sanitary briquettes that can be used in cooking (Markham, 2017). The first step of this project involved building public latrines in some of the poorest areas of Nakura, a city where only 25% of people have access to sanitary facilities. The human waste from these latrines is pumped out and taken to a treatment facility, where it is dried in the sun to reduce its moisture content. The dried waste material is carbonized at high heat, which both sterilizes it and removes the odor. The carbonized material is mixed with a small amount of molasses as a binding agent and is then formed into charcoal briquettes. These briquettes are sold in the community for use in cooking and other applications.

This project is a clear example of how a sustainable solution can provide a net benefit for all three pillars of sustainable development. First, and perhaps most obviously, is the *environmental* benefit. The public latrines provided by this project replace open defecation, which fouls waterways, and the use of pit latrines that are unlined, which fouls groundwater. Eighty percent of the local people currently cook on wood or charcoal, and the production of these fuels places a demand on local wood stocks that is relieved when people cook on briquettes instead. In addition, this project brings *social* benefits to local people. The provision of public latrines provides dignity to those whose only other option would be public defecation. Health is protected because open defecation and fouled water spread disease. And the briquettes produce less smoke than either wood or charcoal, reducing the impact on the health of the users. Of course, this would not be a net benefit if the use of briquettes also spread disease; that is why it is important that the production process reaches high enough heat to kill any pathogens. Finally, there is an *economic* benefit as well. The project employs a number of workers, providing them with jobs and training. The briquettes produced are sold at a price that is lower than the going price for charcoal or wood fuel; and the briquettes are longer burning. In addition to being used in cooking, the briquettes have been used as a consistent source of heat for other uses, including the hatching of chicks. Users report that the consistent and low heat they provide increases the number of chicks who survive, giving an economic benefit to the chick-hatching users (Chebet, 2017).

CREATIVITY AND OPEN-MINDEDNESS WORKING TOGETHER

Keeping in mind that we have seen good reason to prefer focusing on responsibilist intellectual virtues, we can now turn to explore which intellectual character virtues are most relevant to the development and implementation of our particular example of a sustainable solution.

One virtue that might immediately seem relevant in the development of any solutions (sustainable or otherwise) is that of *creativity*.[3] Creativity is required in order to see human waste as a potential resource rather than just as something to be disposed of. This creativity gets the design processes started, but creativity is also used at many steps as those who are engaging in implementation have to work with limited or available resources. As a case in point, the carbonization system used in Nakura was designed using the materials on hand, in this case steel drums. Creativity was required to adapt designs with other materials to make use of the materials that were available.

3. Jason Matteson (2013) has argued for the importance of a virtue of environmental creativity, which I think is probably best classified as a moral virtue. My development here of creativity as an intellectual virtue is not intended to rule out or compete with characterization of creativity as a moral virtue, and particularly as an environmental virtue.

Very often it is this creativity in technical design that we expect in sustainable solutions. But creativity is required in every step of the process, including working out a distribution system for the briquette and even the placement of the public latrines. In addition, creativity is required in the way that the sustainable solution is communicated to the larger community. The importance of creative communication is made clear in this example to help overcome the stigma of using human waste as a fuel source. Designers cannot simply announce their new solution and expect community members to join in. Rather, the designers need to persuasively make the case for their new fuel source, and this will require creativity in the messaging, as well as creativity in the venues and forms through which it is communicated.

Of course, the exercise of the virtue of creativity will often require supplementation by other virtues. Creativity in communication will plausibly require the virtue of patience. As Jason Kawall (Chapter 8, this volume) argues, patience is a virtue that often serves this role; listening to the needs of a community and coming to truly understand their situation will often require a great deal of patience. And this is essential both to developing appropriate sustainable solutions and to communicating their benefits to a specific community. Further creative communication may go beyond just making instrumental arguments that appeal to the needs of a given community. As Susan Clayton demonstrates in Chapter 1 of this volume as well as in her earlier work, an argument illustrating the fairness of a policy might be even more persuasive than one based on the efficacy of the policy in achieving personal or community benefit.

Generally speaking, sustainable solutions will require creativity on the part of the designers and those involved in outreach. However, a focus on the virtue of creativity alone might lead us to look for only technological fixes and the ways that communities can be persuaded to use them. We might think it is appropriate to wait for designers to develop sustainable solutions, and not recognize the active role of the community and its choices in the implementation of sustainable solutions.

Turning to the community, we can see that they need to exhibit open-mindedness in the implementation of sustainable solutions. While this intellectual virtue is required in a wide range of cases, I think that our example highlights how important it can be. Open-mindedness is clearly needed for both the designers and community members to see beyond the stigma of human waste and to be willing to use these briquettes. But open-mindedness is required in other parts of the implementation, too. The location of the public latrines might be a contentious issue, with property owners not wanting to have a latrine nearby. Open-mindedness is required by property owners in order for them to accept the implementation of a program that they might otherwise oppose.

Generally speaking, sustainable solutions require open-mindedness on the part of community members where the program is implemented. However, a focus on the virtue of open-mindedness alone might lead us to look for only lifestyle changes in the community. We should recognize that creativity and open-mindedness work together in the developments and implementation process of sustainable solutions.[4] And, while more creativity or open-mindedness might be required by one group, both designers and community members (who may be the same people) really need both virtues in order to have a reciprocal flow of information and ideas in the design and implementation process.

4. Creativity and open-mindedness might even be seen as two aspects of a more basic virtue like curiosity. Nenad Miščević (2016) has argued that curiosity is the foundational intellectual virtue, since it motivates all inquiry. Clearly, creativity and open-mindedness are often motivated by curiosity, but I prefer to focus on them as two independent virtues here.

Now that we have identified two relevant intellectual virtues, we can turn to examining them in greater detail. Both seem to be intellectual virtues, but are they? An in-depth analysis of each will help us see what makes each a virtue, and it may even help us to further develop our understanding of what intellectual virtues require and how exactly they participate in the development of sustainable solutions.

OPEN-MINDEDNESS AS AN INTELLECTUAL VIRTUE

Open-mindedness is a canonical responsibilist intellectual virtue, appearing on most lists of individual virtues. Even though there are debates about the exact nature of this virtue, there is not a general worry that open-mindedness is either not a virtue or not an intellectual virtue.[5]

Open-mindedness, as characterized on canonical lists of the intellectual virtues, is clearly relevant to implementation of sustainable solutions and particularly in the example that we are exploring. These characterizations focus not only on the behavior that we naturally associate with open-mindedness as a behavior, but also on the underlying psychological mechanisms that help individuals produce that behavior. Wayne Riggs argues that open-mindedness involves being able to see beyond our own "emotions as salience generators" (2015, p. 28). Our emotional reactions are one thing that keep us

5. While the vast majority of explorations of open-mindedness characterize it as a virtue, Jeremy Fantl (2018) has recently argued that there are times that we should not be open-minded, particularly in light of counterarguments that we know to be misleading. Whether his arguments are convincing, they do not undermine open-mindedness in general as an intellectual virtue, but do demonstrate ways in which the virtue of open-mindedness differs from open-mindedness understood as a particular behavior.

from being open-minded, perhaps most clearly in the case of using human waste as fuel. There is clearly a strong (and probably natural) reaction of disgust that would tend to keep us from using this fuel, particularly in cooking. And this emotional reaction is not particularly sensitive to the information that the fuel has been treated to sterilize it from any pathogens. We need to be able to get some distance from our own emotional reactions (including the reaction of disgust) in order to truly "transcend a default cognitive standpoint," as Jason Baehr (2011) thinks is essential in open-mindedness.

Though these accounts of open-mindedness grow out of the literature on intellectual virtues, they illustrate a point that Cheryl Hall makes in Chapter 4 (this volume) on the role of emotional intelligence in developing our virtues. Being able to identify our disgust, observe, it, and evaluate it in a clear-eyed way are all essential parts of overcoming our disgust when open-mindedness calls for it. More generally, being open-minded to sustainable solutions will often require us to let go of old, comfortable patterns of thought and action, asking us to instead take a new and unfamiliar approach. To do this, we need the emotional intelligence to recognize our desire for familiarity, taking a perspective on this emotion that allows us to evaluate a new approach without simply clinging to the comfortable.

Open-mindedness is poised to be a responsibilist intellectual virtue because it is a character trait that can be developed and in being developed reflects well on the individual. We think that being open-minded makes a person a better person overall. And this is in part because being open-minded is something that we can get better at over time, but which requires effort on our part. Of course, individuals may be more or less naturally inclined toward being open-minded; this is a result of our upbringing, environment, and perhaps even genetic predisposition. But no matter how open-minded we are naturally, we can develop and refine our open-mindedness; this is what makes open-mindedness a virtue.

What makes open-mindedness an *intellectual* virtue is its relation to truth. When we act in an open-minded way, but only to curry favor or influence others, this is not an exercise of open-mindedness as a virtue. Intellectual virtues require that we aim at truth. This can require explicitly being motivated by a love of truth, as Linda Zagzebski (1996) requires, or the aiming can be a more implicit part of our intellectual activity in general. But responsibilists generally require that our intellectual virtues aim at truth in some way.[6] Open-mindedness is suited to meet this requirement, as open-mindedness is often a means to discover truths that we would have difficulty coming to realize from our own biased intellectual and emotional perspective. In our example, open-mindedness as an intellectual virtue is required to discover the truth that human waste can be processing into a safe and economically viable fuel source.

CREATIVITY AS AN INTELLECTUAL VIRTUE

Creativity seems like a prima facie good candidate to be an intellectual virtue. The intellectual giants that we admire in our history have generally stood out from the pack by making some discovery, invention, or innovation that was substantially unexpected against the background of their own time. These examples of progress seem to require creativity; we recognize those who suggested radical intellectual leaps far more than those who made workaday progress in a field. Despite this intuitive appeal, creativity is not generally found among lists of the intellectual virtues. This omission is the result of two objections to creativity: that it is not a true virtue but a character trait of another sort, and that it is not a sufficiently intellectual virtue

6. It is a separate question whether intellectual virtues require that a character trait is actually reliable in reaching the truth. See my (2009) and (2010) for a discussion.

because it does not sufficiently aim at intellectual ends, particularly truth. By exploring and responding to both objections, I hope not only to show that creativity is an intellectual virtue, but also to explore its nature as a way of demonstrating that creativity is the kind of intellectual virtue we need to help us implement sustainable solutions.

Objection One: Creativity Is Not a Virtue

One might object that creativity is not a developed trait of character, but rather is a special characteristic possessed by particular outstanding individuals. This objection to creativity being considered a virtue is addressed by Matthew Kieran (2014), who notes that the idea that creativity is a natural gift or talent can be found in philosophers as far back as Plato. This "genius" conception of creativity is supported by the example of outstanding individuals we tend to use to illustrate creativity. The genius model of creativity was also dominant in psychology in the 1950s and 60s, and was reinforced by the work of those like MacKinnon and Barron who studied the differences in backgrounds and personalities between the groups they classed as more and less creative (Barron, 1961; Mackinnon, 1965).[7] Their studies seemed to support the idea that creativity is a result of external factors; as a result, it isn't something that we can develop intentionally through practice and hard work. If creativity is either inborn or a result of one's upbringing, it is not a developed trait of character, but more like a natural endowment. And, as we noted earlier, natural endowments, like eyesight are not virtues on the responsibility model; they are not character traits that reflect well on us.

Linda Zagzebski, in her development of responsibilist intellectual virtue (1996), is also concerned that creativity seems like an outlier in how it is learned. She thinks that it is not learned through

7. For an overview of changing conceptions of creativity, see Amabile and Pillemer (2012).

habituation, as the other Aristotelian virtues are. In fact, she notes that creativity seems diametrically opposed to habit; creative moves are nonhabitual ones. Zagzebski does want to recognize creativity as an intellectual virtue, so she is forced to say that it is importantly different than other virtues, in that even if it can be learned, it is not learned through habituation.

Response to Objection One

In response to these concerns, we might note that recent psychological research has moved away from the genius model. Instead, more recent research has revealed creativity as something that can be intentionally developed over time and which is easier to practice in certain environments. The earlier psychological model was beginning to be challenged even in the 1960s when methods of encouraging creativity were being studied. Sydney Parnes, working with and building on an idea from ad man Alex Osborn, developed techniques of "brainstorming" that are well-known (and perhaps even taken as common sense) today (1967). The central technique they endorsed was that of deferring judgement—generating a list of ideas without evaluating how effective those ideas would be. The development of this technique and others suggests that we are capable of thinking creatively when we want to. Of course, a technique does not yet indicate a developed trait of character. But techniques allow us to practice a skill, and repeating the practice can help us to develop a virtue. If we are looking to develop the virtue of compassion, we might practice the technique of putting ourselves "in the other's shoes." Thus techniques, while not themselves virtues, suggest steps we can take to develop our virtues.

Beyond individual techniques, there is further experimental evidence indicating that we can develop our creativity through practice and training. There are many exercises that have been developed to practice creativity. One example exercise is listing as many alternative uses for a paper clip as possible in a set time period. Repeating exercises like this is a form of training that supplements the techniques discussed earlier. A meta-analysis of studies on creativity training by Scott et. al. found robust evidence that creativity can be developed through training, "Based on 70 prior studies, it was found that well-designed creativity training programs typically induce gains in performance with these effects generalizing across criteria, settings, and target populations" (2004, p. 361). The training programs in these studies varied widely, but the meta-analysis found that a focus on skill development was an effective means to develop creativity. "[M]ore successful programs were likely to focus on development of cognitive skills and the heuristics involved in skill application, using realistic exercises appropriate to the domain at hand" (Scott et al., 2004, p. 361). This finding shows how the development of creativity is like the development of other virtues, in so far as they all have structures relevantly like skills.[8] In response to Zagzebski's concern that creativity cannot be learned through habituation (as other virtues are), we should note that the most effective creativity training programs are those that focus on the repetition of realistic exercises—habituation par excellence.

We can find further evidence that creativity is a virtue when we see it as a mean between extremes. As with open-mindedness, the names of these virtues come from the names we give behaviors. But, being virtues, they cannot be practiced in excess. The same complexity can be seen in other virtues that get their names from common parlance. Courage is a virtue and a mean between cowardice and rashness. We might ordinarily say that someone behaved too courageously in a particular instance; they rushed into fight when they should have merely stood their ground. But it is not possible to

8. For an argument that we ought to treat virtues as a kind of skill see my (2018).

exhibit the virtue of courage in excess; to exhibit excess is to fail to be courageous. Likewise, we could act in ways that we would naturally call too creative. We might get caught up in seeking novelty for its own sake, at the expense of any consideration of practicality or even of taking the time to work through these novel ideas. But this is not to exhibit the virtue of creativity, which requires instead that we learn to avoid both extremes and find a mean. Learning to hit the mean is a process of trial and error, aided by studying other exemplars of the virtue of creativity. Understood as this fine-tuned ability, creativity cannot be an inborn characteristic that only some are lucky enough to receive. Even those for whom creativity is relatively easy must learn how and in what contexts to be creative. Creativity is learned and developed just like the other virtues.

Objection Two: Creativity Is Not an Intellectual Virtue

The second main objection to creativity is against its bona fides as an *intellectual* virtue. Even if creativity can be developed and reflects well on our evaluation, as other virtues do, we might still wonder if it is properly classed as an intellectual virtue. While there is debate about whether intellectual virtues must reach the truth, there is agreement that intellectual virtues generally aim at valuable epistemic states such as truth, knowledge, or understanding. But creativity seems to have a different aim, focusing on producing novel outputs, valuing novelty over accuracy. This feature of creativity motivates a further concern from Zagzebski about the virtue of creativity. Her account of the intellectual virtues requires that they must each be motivated by a love of truth (or knowledge, or cognitive contact with reality). That motivation will take very different forms, leading us to the different intellectual virtues. But creativity seems to aim in a very different direction; a concern with accuracy makes us caution, but creativity throws that caution to the wind, creating new ideas for their own sake. Thus creativity does not meet Zagzebski's criteria for intellectual virtues.

Response to Objection Two

I believe that the simplest way to recognize the truly epistemic ends of particular kinds of creativity is to begin by distinguishing one kind of creativity from the other. I will agree that some kinds of creativity do not have epistemic ends, but also show that there are kinds which are distinctively epistemic.

Let us begin with *artistic creativity*. Examples of artistic creativity are perhaps those that come to mind most readily when we think of exemplars or creativity. Michelangelo, Picasso, and Jackson Pollock are just a few of the visual artists whose generation of new techniques and new styles make them stand out as creative. Similar lists could be generated for creative composers, writers, poets, and so on. When these examples are the first to spring to mind, they make it plausible that creativity is unrelated to truth. Artistic creativity aims at new means of expression, and while we sometimes point to artistic works as getting at a "deeper truth," they are not clearly aimed at generating accurate representations of reality. So (at least without further argument) artistic creativity is not a good candidate to be considered an intellectual virtue.

A second set of examples might be generated when we think about the metaphor of finding ourselves "standing on the shoulders of giants." Copernicus, Galileo, and Newton all stand out as individuals who displayed *theoretical creativity* in the realm of astronomy. We can think of similar examples in a range of sciences. Those who display theoretical creativity do seem to aim at truth, though perhaps in an indirect way. Creativity in the generation of theories tends to generate many false ones; when we aim to create novel theories, those most "far out" from our current best theories will tend to be the least confirmed and most likely to be false. However, theoretical creativity also supports radical scientific discovery. Zagzebski (1996) recognizes that (theoretical) creativity is an intellectual virtue that aims at advancing human knowledge in general, even if it runs the risk of undermining true belief in its possessor. Even though creativity may produce more false beliefs than true ones, and so is unreliable, it is also necessary for human knowledge to progress. Thus, it has a valuable epistemic end. This can be better understood if we think of the creativity as an other-regarding intellectual virtue (Kawall, 2002). Even if creativity does not help its possessor generate more true beliefs than false ones, it does help the overall epistemic community.

The earlier examples of scientific breakthroughs are not yet closely related to ecology or sustainability, but other examples do make that connection. Von Humboldt's discovery that climate and vegetation zones follow a pattern that depends on both latitude and elevation was certainly an exercise in intellectual creativity. Von Humboldt was also the first person to predict climate change and one of the first to highlight the negative effects of monocultures (Wulf, 2015). His intellectual creativity was thus clearly involved in some of the first theorizing about the sustainability of human practices. But ecological theory, no matter how creative, is not sufficient for sustainable solutions, since these require the application of ecological insights to human situations that also consider the social and economic impacts. The development of theory is essential, but sustainable solutions themselves are practice over and above theory. And so theoretical reactivity isn't the best fit to explain the virtue of creativity needed for designing and implementing sustainable solutions.

Turning finally to *practical creativity*, we think of inventors. Examples might include Leonardo da Vinci dreaming up a flying machine or Salk inventing the polio vaccine. This kind of creativity helps us to generate new practical applications of abstract concepts or new solutions to existing problems (rather than new theories per se). We can think of practical creativity as aiming at practical solutions; at changing the world in ways that are beneficial to us. However, changing the world does not seem like a clearly epistemic aim. So, is practical creativity an intellectual virtue? Here I think it is helpful to recognize that true belief and even knowledge are not the only epistemic ends; understanding has been advanced by many as an alternative or complementary epistemic end.[9]

Working toward understanding is a kind of epistemic excellence. And understanding as an epistemic end is a good fit with the responsibilist virtues. While individual actions or instances of belief might aim at truth, we as persons often aim not only to gather a few facts, but also to put those facts together into a deeper understanding.

Once we recognize understanding as an important epistemic aim, we can then see how practical creativity is well poised to be an intellectual virtue motivated by a desire to understand. Practical creativity will require an understanding not only of theories but also of how implementation might be most effective. This requires an understanding of many real-world counterfactuals. For example, if I carbonize this sludge, I will kill the pathogens; if we provide an alternative fuel, deforestation will be reduced. Understanding of each of these counterfactual conditionals is required in order to develop a creative sustainable solution. Theoretical creativity aims to understand the world as it is; practical creativity aims to understand the world as it might be. Thus we can see that even practical creativity should be recognized, not only as a virtue, but also as an intellectual virtue. And so creativity and open-mindedness are two intellectual virtues that are well-suited to help us develop and implement sustainable solutions.

9. Riggs (2003) and Zagzebski (2001) both argue for understanding as a central epistemic aim.

CONCLUSION

Our exploration of the virtues of open-mindedness and creativity have shown us that, despite potential objections, they are in fact intellectual virtues. Why is this conclusion important to the proponent of environmental virtue? It allows for the possibility of integrating open-mindedness and creativity into an overall life of virtue. This means that these virtues, while certainly instrumentally valuable in allowing us to develop and implement sustainable solutions, are also valuable in their own right. We have a reason to develop them outside of their instrumental use. As we develop these virtues, we can also use them in a variety or practical situations to move toward sustainable solutions.

We should recognize that developing and exercising virtues is not easy, and so we might ask ourselves which conditions enable us more easily to act in virtuous ways. Starting with practical creativity, we can note the depth of understanding that is often required for inventing sustainable solutions. Our ecological theories, as well as our theories about the impact of human action on the environment will need to be quite robust in order to allow us to consider the best ways to achieve our sustainability goals and to recognize all the ecological impacts that our actions will have. Thus a call for sustainable solutions must be supplemented by further research into the basic ecological theories undergirding these solutions.

But a strong background of understanding is not sufficient to allow designers to think creatively. As noted earlier, creativity can be learned as a skill, but this requires concentrated practice. One very practical step that can enable designers to develop creativity in their sustainable solutions is to incorporate this creative practice into sustainability education (as suggested by Sandri, 2013). The research on creativity training shows that it works best when it is practiced on problems that are closely related to the target areas of creativity. This gives us a reason not to think of creativity as an all-purpose add-on which can supplement education that is focused on the learning of theories. Rather, educating for creativity should be part of many different fields of study related to sustainable solutions—ecology, economics, systems design, and so on. This conclusion fits with Matt Ferkany's suggestion that we incorporate education for the virtues not only into our general educational systems, but more specifically into education for sustainable development (Ferkany, Chapter 3, this volume). The psychological literature on creativity suggests that this will be the most effective, at least with respect to educating for creativity.

Finally, we should consider the conditions that enable the exercise of the virtue of open-mindedness. As Corral-Verdugo, Frías-Armenta, and Ortiz-Valdez (Chapter 2, this volume) argue, positive social and physical environments encourage the expression of environmental virtues; this is also true of the intellectual virtue of open-mindedness. When a situation seems dire, it is much easier (and often safer) to revert to well-trodden paths and well-known strategies. Open-mindedness requires a degree of stability in one's physical environment, along with trust in those who are presenting a new idea. When sustainable solutions are designed by one group and then presented to another group of end users as a complete package, the solution may not take into account their specific needs and various situations. Accepting such an externally imposed solution may require more open-mindedness (and more sacrifice) than accepting solutions that are the result of participatory design (see Steen, 2013). It will be easier for users to be open-minded to the implementation of sustainable solutions that have taken their needs into account while being developed in a context of open communication and collaborative exploration.[10]

10. Thanks to Jason Kawall for his extensive and constructive comments on an earlier draft of this chapter. Thanks also to the audience at the 2018 Southeastern Epistemology Conference for their feedback, questions, and suggestions.

REFERENCES

Amabile, T.M., and J. Pillemer. 2012. "Perspectives on the social psychology of creativity." *The Journal of Creative Behavior*, *46*(1), 3–15.

Baehr, J. 2011. *The Inquiring Mind: On Intellectual Virtues and Virtue Epistemology.* Oxford: Oxford University Press.

Barron, F. 1961. "Creative vision and expression in writing and painting." In *The Creative Person*, edited by D. W. MacKinnon, 237–251. Berkeley: University of California, Institute of Personality Assessment and Research.

Chebet, C. 2017. "Reaping benefits of human waste energy." *The Standard Digital.* https://www.standardmedia.co.ke/article/2001245439/reaping-benefits-of-human-waste-energy

Code, L. 1987. *Epistemic Responsibility.* Hanover, NH: University Press of New England and Brown University Press.

Fantl, J. 2018. *The Limitations of the Open Mind.* Oxford: Oxford University Press.

Frasz, G. 2005. "Benevolence as an environmental virtue." In *Environmental Virtue Ethics*, edited by R. Sandler and P. Cafaro, 121–134. Lanham, MD: Rowman and Littlefield Publishers.

Greco, J. 2010. *Achieving Knowledge.* Cambridge: Cambridge University Press.

Kawall, J. 2002. "Other-regarding epistemic virtues." *Ratio*, *15*(3), 254–275.

Kawall, J. 2010. "The epistemic demands of environmental virtue." *Journal of Agricultural and Environmental Ethics*, *23*(1–2), 109–128.

Kieran, M. 2014. "Creativity, Virtue and the Challenges from Natural Talent, Ill-Being and Immorality." *Royal Institute of Philosophy Supplement*, *75*, 203–230.

Mackinnon, D.W. 1965. "Personality and the realization of creative potential." *American Psychologist*, *20*, 273–281.

Markham, D. 2017. "In Kenya, dried poop briquettes are serving as a clean cooking fuel." *Treehugger.* https://www.treehugger.com/renewable-energy/kenya-fecal-briquettes-are-serving-clean-cooking-fuel.html

Matteson, J. 2013. "The virtue of environmental creativity." *Environmental Values*, *22*(6), 703–723.

Miščević, N. 2016. "Epistemic value: Curiosity, knowledge and response-dependence." *Croatian Journal of Philosophy*, *16*(3), 393–417.

Parnes, S. 1967. *Creative behavior guidebook.* New York: Scribner's.

Riggs, W. 2003. "Understanding 'virtue' and the virtue of understanding." In *Intellectual Virtue: Perspectives from Ethics and Epistemology*, edited by M. DePaul and L. Zagzebski, 203–226. Oxford: Oxford University Press.

Riggs, W. 2015. "Open-mindedness, insight, and understanding." In *Intellectual Virtues and Education: Essays in Applied Virtue Epistemology*, edited by J. Baehr, 18–37. New York: Routledge.

Roberts, R.C., and W. J. Wood. 2007. *Intellectual Virtues: An Essay in Regulative Epistemology.* Oxford: Oxford University Press.

Sandler, R. 2005. "Introduction." in *Environmental Virtue Ethics,* edited by R. Sandler and P. Cafaro, 1–12. Lanham, MD: Rowman and Littlefield.

Sandri, O.J. 2013. "Exploring the role and value of creativity in education for sustainability." *Environmental Education Research, 19*(6), 765–778.

Scott, G., L.E. Leritz, and M.D. Mumford. 2004. "The effectiveness of creativity training: A quantitative review." *Creativity Research Journal, 16*(4), 361–388.

Sosa, E. 1991. *Knowledge in Perspective.* New York: Cambridge University Press.

Steen, M. 2013. "Virtues in participatory design: Cooperation, curiosity, creativity, empowerment and reflexivity." *Science and Engineering Ethics, 19,* 945–962.

Wright, S. 2009. "The proper structure of the intellectual virtues." *The Southern Journal of Philosophy, 47*(1), 91–112.

Wright, S. 2010. "Internalist virtues and knowledge." *Acta Analytica, 25*(2), 119–132.

Wright, S. 2018. "Are Epistemic Virtues a Kind of Skill?" In *The Routledge Handbook of Virtue Epistemology,* edited by H. Battaly, 58–68. New York: Routledge.

Wulf, A. 2015. *The Invention of Nature: Alexander von Humboldt's New World.* New York: Knopf.

Zagzebski, L. 1996. *Virtues of the Mind.* Cambridge: Cambridge University Press.

Zagzebski, L. 2001. "Recovering understanding." In *Knowledge, Truth, and Duty: Essays on Epistemic Justification, Responsibility, and Virtue,* edited by M. Steup, 235–252. Oxford: Oxford University Press.

INDEX

For the benefit of digital users, indexed terms that span two pages (e.g., 52–53) may, on occasion, appear on only one of those pages.

INDEX

The manufacturer's authorised representative in the EU for product safety is Oxford University Press España S.A. of El Parque Empresarial San Fernando de Henares, Avenida de Castilla, 2 - 28830 Madrid (www.oup.es/en or product.safety@oup.com). OUP España S.A. also acts as importer into Spain of products made by the manufacturer.
Printed and bound by CPI Group (UK) Ltd, Croydon, CR0 4YY
14/09/2026
02216308-0001